CISM COURSES AND LECTURES

The series presents lecture notes, monographs, edited works and proceedings in the field of Mechanics, Engineering, Computer Science and Applied Mathematics.
Purpose of the series is to make known in the international scientific and technical community results obtained in some of the activities organized by CISM, the International Centre for Mechanical Sciences.

INTERNATIONAL CENTRE FOR MECHANICAL SCIENCES

COURSES AND LECTURES - No. 475

ATMOSPHERIC CONVECTION: RESEARCH AND OPERATIONAL FORECASTING ASPECTS

EDITED BY

DARIO B. GIAIOTTI
ARPA FVG OSMER, UDINE, ITALY

REINHOLD STEINACKER
UNIVERSITY OF VIENNA, AUSTRIA

FULVIO STEL
ARPA FVG OSMER, UDINE, ITALY

SpringerWienNewYork

This volume contains 121 illustrations

SPIN 11922483

All contributions have been typeset by the authors.

ISBN-10 3-211-48963-0 SpringerWienNewYork
ISBN-13 978-3-211-48963-5 SpringerWienNewYork

PREFACE

The troposphere is the subtle skin of gases and vapors surrounding our planet which hosts almost all the human activities and weather phenomena. Its name comes from the ceaseless vertical mixing it experiences. The mechanism responsible for the continuous tropospheric vertical mixing is the convection. Even if the physical laws describing convection are quite well known, nevertheless our ability in understanding and forecasting convective weather phenomena is, in some cases, quite poor. The reason for that, depends on our limits in the capability of solving the Navier-Stokes equations, applied to the troposphere, and fully monitoring the actual tropospheric state. These limits are dramatically evident in areas characterized by complex orography and geography, where our knowledge of initial conditions is extremely fragmented and the constraints imposed by the boundaries make the Navier-Stokes equations even more sensitive from those initial conditions. Because of these reasons, when facing aspects related to atmospheric convection, one should reflect on the metaphor by Aldous Huxley, who imagined the relationship between Mankind and Nature as a game of chess where both players know very well the rules, but only Nature never fails neither cheats nor bluffs and the best result Mankind can obtain is a tie. Keeping this in mind, this book is meant to be a collection of information and conceptual models (the rules of the game) on deep-moist atmospheric convection, its related weather phenomena, and it presents several useful tools and techniques (the strategy of the game) not only for the prognostic activities, but even for the forecasts verification, which is fundamental to learn from our mistakes and to develop new techniques. Last but not least, by this book the authors would like to remind to the readers that if the understanding of the atmospheric convection is considered a game, as such, they should enjoy themselves carrying out this difficult, but fascinating and important activity.

In this book the readers are lead along a path that starts with the basics of the atmospheric convection, in particular the dynamics and thermodynamics of unstable moist air. Then the initiation of the spontaneous vertical motions is considered and the weather phenomena related to deep-moist convection are presented. A significant part of the book is devoted to the boundary conditions of convective weather and in particular the role played by orography is examined in detail. The environments hosting convective severe weather are discussed considering their thermic, moisture and wind vorticity fields features, both from theoretical and experimental point of view. The last part of the book is focused on the specific problem of the convective severe weather forecasts and to the evaluation of their quality.

The book is addressed to meteorologists, physicists and weather forecasters, in particular to those interested in improving their knowledge of convection-related weather phenomena. Furthermore the book is considered as a valuable support for the PhD students attending courses on environmental fluid dynamics and meteorology. Each chapter is almost self consistent and there are no propaedeutic parts that the reader should consider necessary before to deal with more advanced ones. Anyway the proposed chapter sequence follows a scheme that the editors consider suitable for the didactics of the topic.
The realization of the book owes much to the scientists who shared with us the CISM course from which it springs from. Special thanks are addressed to the Rectors and the Scientific Editor of CISM who gave the opportunity to organize the course since the beginning; in particular Prof. Manuel Velarde who stimulated us and supported our work throughout its development. Furthermore we are indebted with CISM personnel who helped in preparing the proofs of the book; special thanks to dr. Monica Del Pin who kindly an patiently worked with us.

Dario B. Giaiotti
Reinhold Steinacker
Fulvio Stel

CONTENTS

An Overview of Atmospheric Convection

Paul Markowski

Department of Meteorology, Penn State University, University Park, Pennsylvania, USA

Abstract. The nature of atmospheric convection is briefly reviewed. We can assess the stability of a single air parcel with respect to vertical displacements by comparing the lapse rate of the parcel's environment to the rate of temperature change within the displaced parcel owing to adiabatic expansion or compression and latent heating or chilling. We also can examine the tendency for convective overturning in a global sense, when buoyancy sources are distributed over a large area and the entire fluid is engaged in convective overturning. In this case, the onset of global dry convection due to thermal instability is determined by the Rayleigh number. Within the atmospheric boundary layer on a sunny day, the Rayleigh number is several orders of magnitude larger than the critical Rayleigh number; thus, convective overturning is a ubiquitous characteristic of the atmospheric boundary layer in sunny conditions. The structure of dry atmospheric convection depends to a large degree on the vertical wind shear within the atmospheric boundary layer, and quite possibly also is sensitive to surface characteristics and mean vertical motions.

1 Convection as a means of heat transport

Convection is one of the principal means by which heat is transferred (radiation and conduction are the other two ways). Most generally speaking, it is defined as any fluid motion attributable to the action of a gravitational field upon density variations. In the atmospheric sciences, we typically restrict the use of the term to thermally direct circulations that are driven by unstable vertical distributions of mass. Convection can be viewed as the process used by the atmosphere to accomplish heat transport when conduction and radiation are unable to accomplish the heat transfer quickly enough to prevent an absolutely unstable stratification from arising. Convective motions may be associated with significant deviations from hydrostatic equilibrium. Convection can occur in the absence of condensation or it may lead to condensation, and in certain circumstances, condensation produced by convective overturning can feedback to convective overturning.

In this lecture we will focus on this so-called "buoyant convection."[1] The emphasis is on parcel stability and the shallow version of buoyant convection. Deep moist convection will be the subject of a more detailed investigation in another lecture.

[1]Buoyant convection is sometimes called "free convection," whereas "forced convection" is mechanically driven, often by large wind shear.

2 Parcel stability

Here we will examine the stability of a single parcel of air with respect to vertical displacements in a continuously stratified atmosphere. Before beginning, we must all agree on a suitable definition of a "parcel" of air, the very notion of which can be problematic at times. We often like to think of a parcel of air as having essentially constant variables (e.g., temperature, moisture), but derivatives of variables also are defined for parcels, which implicitly recognizes the presence of variability *within* parcels. So just what does it mean to refer to a volume of air as an "air parcel?" For our purposes herein, we will loosely define an air parcel as a volume of air having a size much smaller than the characteristic scale of the variability of its environment.

Let us begin with a simplified vertical equation of motion which neglects pressure deviations from a hydrostatic resting state, in addition to viscosity and the Coriolis force:

$$\frac{dw}{dt} = B, \tag{2.1}$$

where w is the vertical velocity component and the buoyancy force is approximated as $B = g(T - \bar{T})/\bar{T}$ (the effects of water vapor and liquid water on the density of an air parcel are neglected). Here we regard $\bar{T}$ as the environmental temperature and T as the temperature of an individual air parcel.

Let us displace a parcel vertically from an initial level, z_o, where the parcel is in equilibrium with its environment, i.e., $\bar{T} = T = T_o$, where T_o is the temperature of both the parcel and the environment at the equilibrium position. Because $w = dz/dt$, we can rewrite (2.1) as

$$\frac{d^2\Delta z}{dt^2} = g\,\frac{T - \bar{T}}{\bar{T}}, \tag{2.2}$$

where Δz is the distance of the vertical displacement ($z = z_o + \Delta z$ at some future time; thus, dw/dt can be written as $d^2\Delta z/dt^2$).

The temperature of the displaced parcel at its new position is (using a first-order Taylor series approximation)

$$T \approx T_o - \Gamma_p \Delta z, \tag{2.3}$$

where $\Gamma_p = -\partial T/\partial z$ is the parcel lapse rate. If the parcel is unsaturated during its displacement, $\Gamma_p = \Gamma_d$, the dry adiabatic lapse rate. If the parcel is saturated during its displacement, $\Gamma_p = \Gamma_m$, the moist adiabatic lapse rate.

The temperature of the environment at the new position of the displaced parcel is (using a first-order Taylor series approximation)

$$\bar{T} \approx T_o - \gamma \Delta z, \tag{2.4}$$

where $\gamma = -\partial \bar{T}/\partial z$ is the environmental lapse rate. Therefore, we may write (2.2) as

$$\frac{d^2\Delta z}{dt^2} = -g\,\frac{\Gamma_p - \gamma}{T_o - \gamma\Delta z}\,\Delta z. \tag{2.5}$$

But $T_o \gg \gamma \Delta z$ if the vertical displacement is relatively small; therefore, with reasonable accuracy, we may write (2.5) as

$$\frac{d^2 \Delta z}{dt^2} = -\frac{g}{T_o}(\Gamma_p - \gamma)\ \Delta z \tag{2.6}$$

$$\frac{d^2 \Delta z}{dt^2} + \frac{g}{T_o}(\Gamma_p - \gamma)\ \Delta z = 0. \tag{2.7}$$

Equation (2.7) is a second-order ordinary differential equation having the general solution

$$\Delta z(t) = C_1 e^{i\left[\frac{g}{T_o}(\Gamma_p - \gamma)\right]^{1/2} t} + C_2 e^{-i\left[\frac{g}{T_o}(\Gamma_p - \gamma)\right]^{1/2} t}, \tag{2.8}$$

where C_1 and C_2 are constants that depend on the initial parcel displacement (i.e., the strength of the "push" initially applied to the parcel).

For $\gamma > \Gamma_p$, $i\left[\frac{g}{T_o}(\Gamma_p - \gamma)\right]^{1/2}$ is real and Δz exponentially increases with time—instability is achieved.[2] For $\gamma < \Gamma_p$, $i\left[\frac{g}{T_o}(\Gamma_p - \gamma)\right]^{1/2}$ is imaginary and Δz is oscillatory; i.e., the parcel oscillates about its initial position (z_o) and stability is present.

Conditions for which $\gamma > \Gamma_d$ are said to be "absolutely unstable," and when $\gamma < \Gamma_m$, conditions are said to be "absolutely stable." When $\Gamma_m < \gamma < \Gamma_d$, the atmosphere is "conditionally unstable" (stable with respect to unsaturated vertical displacements, unstable with respect to saturated vertical displacements). When $\gamma = \Gamma_d$ ($\gamma = \Gamma_m$) the atmosphere is said to be "neutral" with respect to dry (saturated) vertical displacements. Lastly, when $\gamma_m > \Gamma_m$, where $\gamma = \gamma_m$ when the environmental lapse rate is saturated, the atmosphere is regarded as "moist absolutely unstable."

In the above idealization, we have neglected the perturbation pressures that arise from the departures from hydrostatic balance (actually, we have neglected these perturbations *twice*—once in the vertical momentum equation, and once in the approximation for buoyancy). In subsequent lectures it will be shown that this approximation loses credibility for more realistic models of convection, and in fact, in severe convective storms, such pressure fluctuations play a critical dynamical role.

3 Stability as defined by available energy

There is often confusion between the lapse rate definition of stability (reviewed in the previous section), which depends on the local lapse rate compared to the dry and moist adiabatic lapse rates, and what sometimes is referred to as the *available-energy* definition of stability, which depends on whether a parcel possesses positive buoyant energy. For example, a parcel will accelerate upward if it is warmer than its surroundings, even if the

[2] The initial condition that $\Delta z = 0$ at $t = 0$ requires that $C_1 = -C_2$; therefore, C_2 cannot equal zero (which would preclude exponential growth of Δz in time when $\gamma > \Gamma_p$) unless C_1 also equals zero, which happens to be possible only if the additional initial condition is that $d\Delta z/dt = 0$ at $t = 0$. This additional initial condition is not possible if a parcel is displaced vertically; thus, we can be assured that any vertical displacement exponentially grows when $\gamma > \Gamma_p$.

environmental lapse rate at that level is less than the moist adiabatic lapse rate. Consider an environment containing significant convective available potential energy (CAPE). By definition, rising parcels of air will be positively buoyant and will be subjected to an upward acceleration at any level between the level of free convection (LFC) and the equilibrium level (EL). But the lapse rate of the environmental temperature profile between the LFC and EL may vary—and may even be "absolutely stable" in some layers according to the lapse rate definition of stability. The concept of buoyancy and the available energy definition of stability will be expanded upon in another lecture.

4 Global dry convection

4.1 The Rayleigh number

We'll next turn our attention to the onset and characteristics of convective overturning when buoyancy forces are distributed over a large horizontal area compared to the depth of the overturning layer, i.e., when the buoyancy forcing is global, not local, in nature. The conditions for which convective overturning commences first were studied roughly 100 years ago, both theoretically and experimentally, by Rayleigh and Benard, respectively.[3] The onset of convection in a thin layer of fluid heated from below commences when a nondimensional number, called the Rayleigh number, exceeded a critical threshold. The Rayleigh number, Ra, is expressed as

$$\mathrm{Ra} = \frac{g\ \beta\ \Delta T\ H^3}{\kappa \nu}, \tag{4.1}$$

where κ and ν are the heat and momentum diffusivities, respectively, β is the coefficient of thermal expansion $[= \frac{1}{\alpha}\left(\frac{\partial \alpha}{\partial T}\right)_p$ for air, where α is the specific volume, p is pressure, and T is temperature], and ΔT is the temperature difference between two plates separated by a vertical distance H. As the temperature differential, ΔT, increases, Ra also increases. But diffusion has a stabilizing effect; thus, as κ and ν increase, Ra decreases.

When $\mathrm{Ra} > \mathrm{Ra}_c$, where Ra_c is the critical Rayleigh number, heating of the fluid from below exceeds the ability of the fluid to transfer heat by molecular conduction; thus, convective overturning ensues. It can be shown after much tedious calculus that the minimum Ra_c that must be exceeded for convective overturning to commence is approximately 120. This Ra_c occurs when the upper and lower boundaries are free-slip and perfectly insulating and there is no mean vertical wind shear. The value of Ra_c increases if mean vertical wind shear or conducting or no-slip boundaries are imposed; i.e., these effects have a suppressing effect on the tendency for convective overturning. In sunny, terrestrial conditions, the atmospheric boundary layer easily satisfies $\mathrm{Ra} > \mathrm{Ra}_c$. Once convection commences, one might naturally wonder how the overturning motions are organized, if at all.

[3] It is now known that Benard's experiments did not actually demonstrate the same thermal instability studied by Rayleigh. Because of the thinness of the fluid used, Benard instead discovered an instability that arises due to surface tension differences that arise when a fluid is unevenly heated.

4.2 Organization of convective motions

Convective motions in the atmospheric boundary layer often have a strong *nonrandom* element, which may be somewhat surprising since randomness is one of the defining characteristics of turbulence. Convective motions often are organized into hexagonal cellular patterns, especially when the vertical wind shear is weak or absent. There is no rigorous theory explaining why hexagonal cells should be favored over other geometries such as rectangles or triangles, but some have hypothesized that hexagons are favored because nature seeks a geometrical configuration that minimizes the amount of "fence-building" between cells, i.e., one that minimizes the perimeter-to-area ratio.[4] The horizontal wavelength of the convective cells, based on theory, is roughly three times the depth of the overturning layer. Observations of cells in atmospheric boundary layers, however, have revealed much wider cells, with width-to-height ratios ranging from 10–30:1. "Open" cellular convection is characterized by walls of cloud surrounding open, cloudless areas, with descent in these central open areas. "Closed" cellular convection comprises rings of open areas surrounding solid cloud, with apparently upward motion in the center (where the cloud is) and sinking motion in the rings. Cellular convection is most commonly observed over the oceans, and there is some suggestion that open (closed) cells tend to be favored over relatively warm (cold) water. Some laboratory studies of convection also suggest that open (closed) cellular convection depends on the large-scale vertical motion, with open (closed) cellular convection favored when the large-scale vertical velocity is downward (upward).

Convective motions in the atmospheric boundary layer also can be organized into lines of counterrotating horizontal vortices, called horizontal convective rolls (HCRs), particularly when the vertical wind shear is significant. Because of the suppressing effect of mean vertical wind shear on convective overturning due to thermal instability, HCRs are often aligned with the mean vertical wind shear in the boundary layer, thereby minimizing the component of the wind shear that lies in the plane of the convective motions. However, the presence of mean wind shear also can be associated with dynamical instabilities (e.g., Kelvin-Helmholtz instability), which promotes roll orientation normal to the shear. Thus, the superposition of thermal instability with dynamical instability can lead to roll orientations that deviate from alignment with the mean shear. HCRs can give rise to the familiar "cloud streets" or nearly unbroken cloud bands within the boundary layer, if ample moisture is present in the boundary layer. The ascending branch of the roll circulations is the favored area for the formation of clouds, and, as was the case for open and closed cells, the aspect ratio of the roll circulations is proportional to the depth of the boundary layer, with spacings between updrafts of 3–10 times the depth of the overturning layer typically observed.

5 Further reading

Agee, E. M., T. S. Chen, and K. E. Dowell, 1973: A review of mesoscale cellular convection. *Bull. Amer. Meteor. Soc.*, **54**, 1004–1011.

[4] A circle has the smallest perimeter-to-area ratio. The hexagon is the polygon that most closely approximates a circle and can be fitted into an array of elements with common boundaries.

Bryan, G. H., and M. J. Fritsch, 2000: Moist absolute instability: The sixth static stability state. *Bull. Amer. Meteor. Soc.*, **81**, 1207–1230.

Emanuel, K. A., 1994: *Atmospheric Convection.* Oxford University Press, 580 pp.

Hess, S. L., 1959: *Introduction to Theoretical Meteorology.* Holt, 362 pp.

Saucier, W. J., 1955: *Principles of Meteorological Analysis.* Dover, 438 pp.

Schultz, D. M., P. N. Schumacher, P. N., and C. A. Doswell, 2000: The intricacies of instabilities. *Mon. Wea. Rev.*, **128**, 4143–4148.

Sherwood, S. C., 2000: On moist instability. *Mon. Wea. Rev.*, **128**, 4139–4142.

Weckwerth, T. M., J. W. Wilson, R. M. Wakimoto, and N. A. Crook, 1997: Horizontal convective rolls: Determining the environmental conditions supporting their existence and characteristics. *Mon. Wea. Rev.*, **125**, 505–526.

Weckwerth, T. M., T. W. Horst, and J. W. Wilson, 1999: An observational study of the evolution of horizontal convective rolls. *Mon. Wea. Rev.*, **127**, 2160–2179.

The Concept of Buoyancy and Its Application to Deep Moist Convection

Paul Markowski

Department of Meteorology, Penn State University, University Park, Pennsylvania, USA

Abstract. The origins of the buoyancy force and the vertical perturbation pressure gradient force are reviewed. A discussion of parcel theory and its limitations follows. The quantitative determination of updraft velocity from environmental soundings is complicated by factors such as environmental heterogeneity, hydrometeor loading, freezing, entrainment, and vertical perturbation pressure gradients.

1 The concept of buoyancy

It seems intuitive that warm air should rise and cold air should sink. We routinely attribute these motions to "buoyancy forces." But where are the so-called buoyancy forces in the vertical momentum equation, which, in its inviscid form, is

$$\frac{dw}{dt} = -\frac{1}{\rho}\frac{\partial p}{\partial z} - g, \tag{1.1}$$

where w is the vertical velocity component, ρ is the air density, p is the pressure, and g is the gravitational acceleration? The answer is that they are implicitly contained in both terms on the right hand side of (1.1).

Buoyancy is an unbalanced vertical pressure gradient force attributable to variations in density within the atmospheric column. To obtain a "buoyancy force," let us rewrite (1.1) as

$$\rho\frac{dw}{dt} = -\frac{\partial p}{\partial z} - \rho g, \tag{1.2}$$

and define a horizontally homogeneous base state pressure and density field (denoted by overbars) that is in hydrostatic balance, such that

$$0 = -\frac{\partial \bar{p}}{\partial z} - \bar{\rho} g. \tag{1.3}$$

Subtracting (1.3) from (1.2) yields

$$\rho\frac{dw}{dt} = -\frac{\partial p'}{\partial z} - \rho' g, \tag{1.4}$$

where primed p and ρ variables are the deviations of the pressure and density field from the horizontally homogeneous, balanced base state. Rearrangement of terms in (1.4) yields

$$\frac{dw}{dt} = -\frac{1}{\rho}\frac{\partial p'}{\partial z} - \frac{\rho'}{\rho}g \tag{1.5}$$

$$= -\frac{1}{\rho}\frac{\partial p'}{\partial z} + B \tag{1.6}$$

where $B = -\frac{\rho'}{\rho}g$ is the familiar buoyancy force and $-\frac{1}{\rho}\frac{\partial p'}{\partial z}$ is the vertical perturbation pressure gradient force (VPPGF). The VPPGF arises from velocity gradients (which imply *accelerations* of air) and from the presence of density anomalies. The VPPGF therefore has both nonhydrostatic and hydrostatic components. The VPPGF will be treated only qualitatively in this lecture. A more thorough decomposition of the VPPGF will be undertaken in another lecture.

It is common for the buoyancy force to be written as

$$B = -\frac{\rho'}{\bar{\rho}}g, \tag{1.7}$$

where $\rho = \rho(x, y, z, t)$ has been replaced by $\bar{\rho} = \bar{\rho}(z)$ in the denominator (a "base state" density profile). This is the equivalent of making the anelastic approximation. Using the equation of state and neglecting products of perturbations, it can be shown that

$$\frac{\rho'}{\bar{\rho}} \approx \frac{p'}{\bar{p}} - \frac{T_v'}{\bar{T}_v} \approx -\frac{T_v'}{\bar{T}_v}, \tag{1.8}$$

where T_v is the virtual temperature and we have assumed that $|p'/\bar{p}| \ll |T_v'/\bar{T}_v|$. The buoyancy force therefore can be approximated as

$$B \approx \frac{T_v'}{\bar{T}_v}g. \tag{1.9}$$

Often it is customary to designate the "base state" virtual temperature as that of the ambient environment, and the perturbation virtual temperature as the temperature difference between the environment and an air parcel rising through an updraft, so that

$$B \approx \frac{T_{v_p} - T_{v_{env}}}{T_{v_{env}}}g, \tag{1.10}$$

where T_{v_p} is the virtual temperature of an air parcel rising through the updraft[1] and $T_{v_{env}}$ is the virtual temperature of the ambient environment. When an air parcel is warmer than the environment, a positive buoyancy force exists, resulting in upward acceleration. Although the terms "base state" and "environment" often are used interchangeably, I

[1] Here and hereafter the discussion often only mentions updrafts, but the same arguments also can be applied to downdrafts.

personally do not favor such usage. It is my opinion that "base state" should refer to something that does not change in time. Clearly, the ambient environment of a convective cloud unavoidably *does* change in time, due to compensating motions. If we constantly need to be concerned about ever-changing base states, then I feel that the decomposition of variables into perturbation and base state terms loses its advantages. Therefore, I believe that the base state atmospheric conditions should refer to those conditions characterizing the environment prior to the development of convection,[2] and the base state quantities should be held fixed in time. [Based on the argument I've put forth, I obviously would advocate the use of (1.9) over (1.10).]

I think that meteorologists have latched on the concept of buoyancy because it facilitates understanding of why air should accelerate upward or downward in certain circumstances. For example, it is not obvious why warm air should rise when inspecting (1.1). But after defining a horizontally homogeneous, hydrostatic base state and considering deviations from this base state in the derivation of (1.6) and (1.9), it is quite apparent that warm air should rise indeed.

2 Parcel theory

The simplest approach to predicting the vertical velocity of a convective element, called "parcel theory," is based on assuming that

$$\frac{dw}{dt} = B. \tag{2.1}$$

If we multiply both sides of (2.1) by $w \equiv dz/dt$, we can obtain

$$w\frac{dw}{dt} = B\frac{dz}{dt} \tag{2.2}$$

$$\frac{d}{dt}\left(\frac{w^2}{2}\right) = B\frac{dz}{dt} \tag{2.3}$$

$$dw^2 = 2B\,dz. \tag{2.4}$$

Next, we integrate (2.4) from the level of free convection (LFC) to the equilibrium level (EL). We will assume that $w = 0$ at the LFC, since the only force considered here is the buoyancy force (which does not become positive until above the LFC, by definition). Also, we will assume that the maximum vertical velocity occurs at the EL (also a justifiable assumption given that only the buoyancy force has been considered). Integration of (2.4)

[2] Another issue is how one goes about defining a representative base state, given the tremendous horizontal variability often present on the mesoscale. This issue will have to be avoided at the present time in the interest of brevity.

yields

$$\int_{LFC}^{EL} dw^2 = 2\int_{LFC}^{EL} B\,dz \tag{2.5}$$

$$w_{EL}^2 - w_{LFC}^2 = 2\int_{LFC}^{EL} B\,dz \tag{2.6}$$

$$w_{max}^2 = 2\int_{LFC}^{EL} B\,dz \tag{2.7}$$

$$w_{max} = \sqrt{2\,\mathrm{CAPE}}, \tag{2.8}$$

where CAPE is the convective available potential energy defined by

$$\mathrm{CAPE} = \int_{LFC}^{EL} B\,dz. \tag{2.9}$$

The prediction of w_{max} in a convective updraft by (2.8) typically is too large using parcel theory, for several reasons: (1) vertical perturbation pressure gradients (sometimes called "aerodynamic drag") were neglected; (2) mixing with ambient environmental air was neglected (typically B is computed assuming undiluted, moist adiabatic ascent); (3) compensating subsidence within the surrounding air (which can affect B and/or the perturbation pressure field) was ignored; (4) the weight of condensate within a parcel rising moist adiabatically was ignored. Therefore, the value of w_{max} predicted by (2.8) can be interpreted as an upper limit for vertical velocity in buoyant convection. In fact, this w_{max} sometimes is called the "thermodynamic speed limit."

3 Deviations from parcel theory

3.1 Vertical perturbation pressure gradients

Although the concept of buoyancy has been useful in developing our understanding of the behavior of convective phenomena, the concept sometimes carries some undesirable baggage with it. For example, one might ask, "how do the parcels in the middle of an updraft, shielded from the ambient environment, 'know' that they are warmer than the ambient environment and supposed to rise?" This is a perfectly legitimate question, although it arises only because we have introduced the notion of buoyancy forces. Parcels really don't need to "know" that they are warmer or colder than their surroundings. This "knowledge" is implicitly taken care of by the righthand side of (1.1). Parcels really only care about pressure gradients, and how these gradients are balanced or unbalanced by gravity.

Parcel theory traditionally neglects vertical perturbation pressure gradients. Yet these are vitally important, for they tend to offset some of the acceleration induced by the buoyancy force.[3] Relatively high (low) pressure tends to be located above a rising (sinking) warm (cold) bubble, and relatively low (high) pressure tends to be located beneath a rising (sinking) warm (cold) bubble. Thus, an upward-directed buoyancy force (associated with a warm bubble) tends to be associated with a downward-directed perturbation pressure gradient force, and a downward-directed buoyancy force (associated with a cold bubble) tends to be associated with an upward-directed perturbation pressure gradient force. One physical explanation for such perturbation pressures and their gradients is that a positive perturbation pressure (relatively high pressure) must exist above a rising bubble in order to push air laterally out of the way of the rising bubble, and that a negative perturbation pressure (relatively low pressure) must exist beneath a rising bubble in order to draw air into the wake of the rising bubble, in order to preserve mass continuity. Conversely, a cold bubble tends to have relatively high (low) pressure beneath (above) it for the same reasons. Furthermore, the presence of a temperature anomaly alone, regardless of whether or not it is rising or sinking, leads to pressure perturbations, owing to the fact that temperature anomalies are associated with thickness changes; i.e., pressure surfaces are perturbed by temperature anomalies (thereby giving rise to pressure anomalies) in an attempt to preserve hydrostatic balance. In short, when considering the effect of the perturbation pressure gradient, warm (cold) bubbles tend not to rise (sink) as fast as one would expect based on the consideration of the buoyancy force alone (Fig. 1).

If the cold or warm anomaly is narrow, then the buoyancy force is larger in magnitude than the perturbation pressure gradient forces, and warm (cold) air does in fact rise (sink). However, as a warm (cold) bubble increases in width, more air must be pushed out of its way in order for it to rise (sink), and more air must be drawn in below (above) to compensate for the wider region of ascent (descent). Thus, the opposing perturbation pressure gradient increases in magnitude with respect to the buoyancy force as a warm or cold bubble increases in width.

When a warm or cold bubble becomes very wide, the opposing vertical perturbation pressure gradient becomes so large that it entirely offsets the buoyancy force, and the net acceleration is *zero*. This is the hydrostatic limit, in other words, the horizontal length scale of the temperature anomaly has become very large compared to the vertical scale—the vertical pressure gradient and gravity are in balance.

This discussion of perturbation pressure gradients began when we posed the question "how do parcels in the interior of updrafts 'know' that they are warmer than their environments?" As shown above, buoyancy-driven vertical accelerations increase as updrafts/downdrafts narrow. This is just another way of saying that parcels in skinny updrafts/downdrafts are closer to the ambient environment than they are in a wide updraft (because they are less shielded in a skinny updraft/downdraft than they would be in a wide updraft/downdraft). Because they are "closer" to the ambient environment, the updraft (downdraft) parcels better "know" that they are warmer (colder) than the

[3] An exception is for updrafts occurring in environments containing large vertical wind shear, in which the perturbation pressure gradient force may act in the same direction as buoyancy, especially at low levels, thereby augmenting the vertical acceleration. This effect will be discussed in greater detail in another lecture.

ambient environment and are supposed to rise (sink). When an updraft (downdraft) becomes very wide, we can imagine that interior parcels are so distant from the ambient environment that they cannot "know" that they are warmer (colder) than the ambient environment; thus, these interior parcels do not rise (sink), and theoretically, this result was attributed to the size of the opposing vertical perturbation pressure gradients becoming so large that they opposed buoyancy altogether (the hydrostatic limit).

In summary, parcel accelerations are related to the sum of buoyancy and perturbation pressure gradient forces. Perturbation pressure gradient forces tend to oppose the buoyancy force (at least in the absence of strong shear, which we will not consider in this lecture), so that the net vertical accelerations predicted by parcel theory, which neglects perturbation pressure gradients, tend to be larger than those in reality. And when a warm or cold anomaly becomes very wide, the opposing vertical pressure gradient force can entirely balance the buoyancy force, resulting in no net vertical acceleration.

3.2 Entrainment

Parcel theory also neglects the exchange of momentum, moisture, and temperature between the parcel and its environment. Mixing of environmental air into a rising air parcel typically slows the parcel by reducing its buoyancy and upward momentum. This process is called entrainment. Entrainment can be viewed as a parcel dilution process, because the equivalent potential temperature (θ_e) of a rising parcel typically is reduced by entrainment, leading to the realization of less CAPE and smaller w_{max} than predicted by parcel theory (Fig. 2).

Updraft dilution increases with the tilt of an updraft, which increases the surface area of the updraft that is exposed to the hostile ambient environment. The entrainment into the sides of an updraft also increases as the vertical acceleration within the updraft increases, owing to mass continuity. (This sometimes is referred to as "dynamic entrainment.") As updraft width increases, the core of the updraft can become better shielded from the effects of entrainment. For this reason, skinny updrafts are more susceptible to the detrimental effects of entrainment than are wide updrafts. Updraft (and often cloud) widths tend to increase in regions where strong mesoscale ascent is occurring, which might be why such regions of ascent are most favorable for the initiation and maintenance of thunderstorms.

In terms of quantifying entrainment, it can be shown that the time rate of change of the temperature "excess" (how warm a parcel is compared to its environment) of a cloudy, buoyant parcel can be reasonably approximated as

$$\frac{dT_v'}{dt} = (\gamma - \Gamma_m)w - \left[T_v' + \frac{L}{c_p}q'\right]\lambda, \tag{3.1}$$

where T_v' and q' are the virtual temperature and specific humidity excesses over the ambient environment, w is the vertical velocity, γ is the environmental lapse rate, Γ_m is the moist adiabatic lapse rate, L is the latent heat of vaporization, c_p is the specific heat at constant pressure, and λ is the entrainment rate. It is possible to estimate the entrainment rate from in situ thermodynamic measurements within a cloud. In simple one-dimensional cloud models, the entrainment rate often is parameterized in terms of updraft width.

3.3 Hydrometeor loading

As air parcels rise moist adiabatically, condensate is carried along within the parcels (for pseudoadiabatic ascent, it is assumed that the condensate is *not* carried along). The condensate has mass, and the weight of the condensate exerts a downward acceleration equal to gq_c, where q_c is the mass of condensate per kg of air (maximum values of q_c within a strong updraft typically are 8–18 g kg^{-1}). The effect of condensate or hydrometeor "loading" on a parcel may be incorporated into the buoyancy term, e.g., we may redefine B as

$$B = g\left(\frac{T_v'}{\bar{T}_v} - q_c\right). \tag{3.2}$$

Thus, a 3 K virtual temperature excess will be offset entirely (assuming $\bar{T}_v \sim 300$ K) by a condensate concentration of 10 g kg^{-1}.

3.4 Freezing and melting

The latent heat of fusion contributes to parcel buoyancy as well. Freezing of water droplets at high altitudes within updrafts contributes to positive buoyancy. Melting of hail, snow, and graupel within descending precipitation shafts contributes to negative buoyancy. Freezing and melting effects on temperature and buoyancy are not included in the calculations of the pseudoadiabats displayed on a thermodynamic diagram. Freezing effects have been found to provide a small extra "boost" to updrafts above the freezing level, and melting effects have been found to be important in the generation of downdrafts in convective storms.

4 Further reading

Das, P., 1979: A non-Archimedian approach to the equations of convection dynamics. *J. Atmos. Sci.*, **36**, 2183–2190.

Doswell, C. A., and P. M. Markowski, 2004: Is buoyancy a relative quantity? *Mon. Wea. Rev.*, **132**, 853–863.

Houze, R. A., 1993: *Cloud Dynamics.* Academic Press, 573 pp.

List, R. and E. P. Lozowski, 1970: Pressure perturbations and buoyancy in convective clouds. *J. Atmos. Sci.*, **27**, 168–170.

Warner, J., 1970: On steady-state one-dimensional models of cumulus convection. *J. Atmos. Sci.*, **27**, 1035–1040.

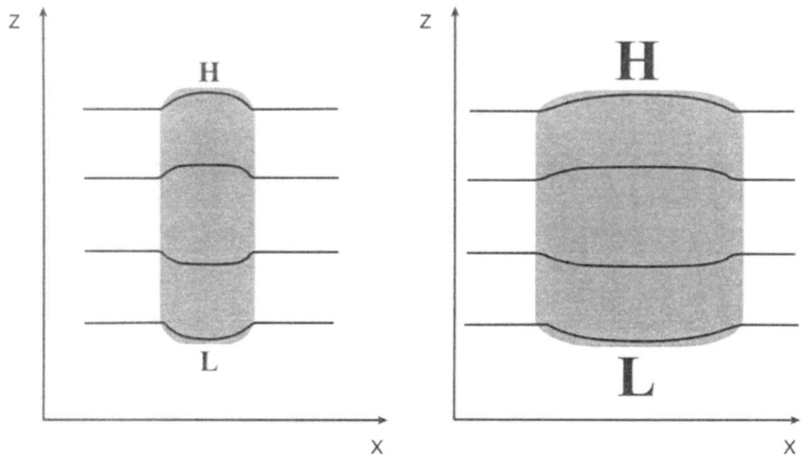

Figure 1. Parcel theory neglects the effect of pressure perturbations on a rising air parcel. This effect is partly due to the presence of the temperature (density) anomaly in the column, and partly due to so-called "dynamic effects"—the need to evacuate fluid ahead of the rising thermal and replace fluid in the wake of the thermal. When two warm perturbations of the same amplitude (i.e., same temperature excess) but differing widths are introduced, the upward-directed buoyancy forces are identical, but the downward-directed pressure gradient force increases as the width of the thermal increases. Thus, the net acceleration upward is larger in the case on the left, because the thermal anomaly is narrower. Black lines are pressure surfaces (note that they buckle upward and downward within the warm anomaly, in accordance with the fact that the thickness is larger within the warm anomaly).

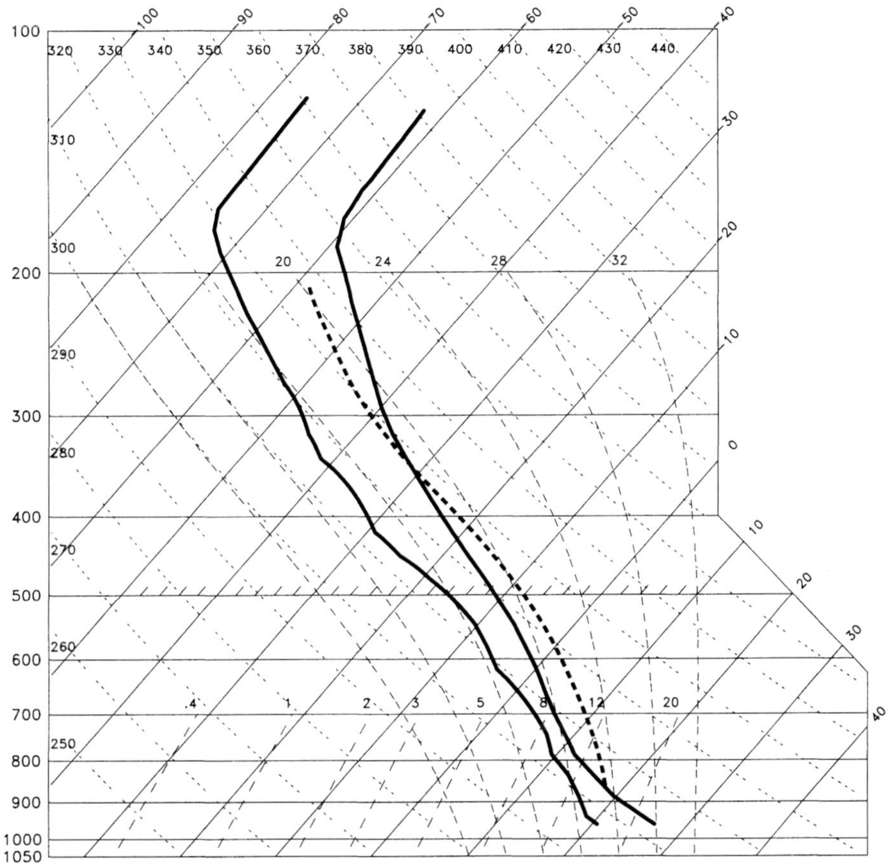

Figure 2. The effect of entrainment on the trajectory that an updraft parcel might follow on a skew T-log p diagram.

Pressure Fluctuations Associated with Deep Moist Convection

Paul Markowski

Department of Meteorology, Penn State University, University Park, Pennsylvania, USA

Abstract. The momentum equations can be written in a form whereby pressure is replaced by a deviation of pressure from a hydrostatic, horizontally homogeneous base state. This pressure perturbation has both hydrostatic and nonhydrostatic parts owing to density deviations from the horizontally homogeneous base state and three-dimensional wind velocity gradients, respectively. The pressure perturbation also can be decomposed into what are referred to as a dynamic pressure perturbation and a buoyancy pressure perturbation. It is shown that dynamic pressure perturbations arise when deformation or vorticity are present in the velocity field. Buoyancy pressure perturbations arise when vertical buoyancy gradients exist.

1 Review of the origins of pressure perturbations

Let us describe the total pressure p and density ρ as the sum of a horizontally homogeneous base state pressure and density and a deviation from this base state, i.e.,

$$p(x,y,z,t) = \bar{p}(z) + p'(x,y,z,t) \tag{1.1}$$

$$\rho(x,y,z,t) = \bar{\rho}(z) + \rho'(x,y,z,t), \tag{1.2}$$

where the base state is denoted with overbars, the deviation from the base state is denoted with primes, and the base state is defined such that it is in hydrostatic balance ($\frac{\partial \bar{p}}{\partial z} = -\bar{\rho} g$). The inviscid vertical momentum equation can then be written as

$$\frac{dw}{dt} = -\frac{1}{\rho}\frac{\partial p'}{\partial z} - \frac{\rho'}{\rho} g, \tag{1.3}$$

where w is the vertical velocity component, g is the gravitational acceleration, and the vertical component of the Coriolis force has been neglected. The forces on the rhs of (1.3) are the vertical perturbation pressure gradient and buoyancy, respectively. In this lecture we will investigate further the nature of pressure perturbations.

2 Hydrostatic and nonhydrostatic pressure perturbations

The perturbation pressure, p', can be represented as the sum of a hydrostatic pressure perturbation p'_h and a nonhydrostatic pressure perturbation p'_{nh}, i.e.,

$$p' = p'_{nh} + p'_h. \tag{2.1}$$

The former arises from density perturbations by way of the relation

$$\frac{\partial p'_h}{\partial z} = -\rho' g, \tag{2.2}$$

which allows us to rewrite (1.3) as

$$\frac{dw}{dt} = -\frac{1}{\rho}\frac{\partial p'_{nh}}{\partial z}. \tag{2.3}$$

Hydrostatic pressure perturbations occur beneath buoyant updrafts (where $p'_h < 0$) and within the latently chilled precipitation regions of convective storms (where $p'_h > 0$). The nonhydrostatic pressure perturbation is simply the difference between the total pressure perturbation and hydrostatic pressure perturbation. It tends to be associated with accelerations of air, which themselves are effected by hydrostatic pressure perturbations. A more descriptive explanation of such pressure perturbations is provided below.

3 Dynamic and buoyancy pressure perturbations

Another common approach undertaken to decompose the perturbation pressure is to form a diagnostic pressure equation by taking the divergence ($\boldsymbol{\nabla}\cdot$) of the three-dimensional momentum equation,

$$\frac{\partial \mathbf{v}}{\partial t} + \mathbf{v}\cdot\boldsymbol{\nabla}\mathbf{v} = -\alpha_o \boldsymbol{\nabla} p' + B\mathbf{k} - f\mathbf{k}\times\mathbf{v} \tag{3.1}$$

where $\mathbf{v} = (u, v, w)$ is the velocity vector, $\alpha_o \equiv 1/\rho_o$ is a constant specific volume, $B = -\rho' g/\rho$ is the buoyancy, and f is the Coriolis parameter (the Coriolis force, $-2\boldsymbol{\Omega}\times\mathbf{v}$, where $\boldsymbol{\Omega}$ is the earth's angular velocity, has been approximated as $-f\mathbf{k}\times\mathbf{v}$). Here the Boussinesq momentum equation is used for simplicity; the use of the fully compressible momentum equations results in a few additional terms, but these would not severely impact a qualitative assessment of the origins of pressure perturbations.

The divergence of (3.1) is

$$\frac{\partial(\boldsymbol{\nabla}\cdot\mathbf{v})}{\partial t} + \boldsymbol{\nabla}\cdot(\mathbf{v}\cdot\boldsymbol{\nabla}\mathbf{v}) = -\alpha_o\nabla^2 p' + \frac{\partial B}{\partial z} - \boldsymbol{\nabla}\cdot(f\mathbf{k}\times\mathbf{v}). \tag{3.2}$$

Using $\boldsymbol{\nabla}\cdot\mathbf{v} = 0$, we obtain

$$\alpha_o\nabla^2 p' = -\boldsymbol{\nabla}\cdot(\mathbf{v}\cdot\boldsymbol{\nabla}\mathbf{v}) + \frac{\partial B}{\partial z} - \boldsymbol{\nabla}\cdot(f\mathbf{k}\times\mathbf{v}). \tag{3.3}$$

After evaluating $\boldsymbol{\nabla}\cdot(\mathbf{v}\cdot\boldsymbol{\nabla}\mathbf{v})$ and $\boldsymbol{\nabla}\cdot(f\mathbf{k}\times\mathbf{v})$, we obtain

$$\begin{aligned}\alpha_o\nabla^2 p' \;=\; & -\left[\left(\frac{\partial u}{\partial x}\right)^2 + \left(\frac{\partial v}{\partial y}\right)^2 + \left(\frac{\partial w}{\partial z}\right)^2\right] \\ & -2\left(\frac{\partial v}{\partial x}\frac{\partial u}{\partial y} + \frac{\partial w}{\partial x}\frac{\partial u}{\partial z} + \frac{\partial w}{\partial y}\frac{\partial v}{\partial z}\right) + \frac{\partial B}{\partial z} + f\zeta - \beta u,\end{aligned} \tag{3.4}$$

where $\zeta = \frac{\partial v}{\partial x} - \frac{\partial u}{\partial y}$ and $\beta = df/dy$. The last term on the rhs of (3.5) is associated with the so-called β-effect and is small, even on the synoptic scale. The second-to-last term on the rhs of (3.5) is associated with the Coriolis force. The remaining terms will be discussed momentarily.

On the synoptic scale, the Coriolis force tends to dominate (3.5) and, neglecting the β-effect, we obtain

$$\alpha_o \nabla^2 p' = f\zeta. \tag{3.5}$$

When p' is reasonably "well-behaved," $\nabla^2 p' \propto -p'$; therefore,

$$p' \propto -f\zeta, \tag{3.6}$$

which is the familiar synoptic-scale relationship between pressure perturbations and flow curvature: anticyclonic flow is associated with high pressure and cyclonic flow is associated with low pressure.

Hereafter we will neglect the terms in (3.5) associated with the Coriolis force and β-effect. Also, it will be helpful to rewrite (3.5) in terms of vorticity ($\boldsymbol{\omega}$) and deformation ($\mathbf{D}$) vectors, where

$$\boldsymbol{\omega} \equiv \left(\frac{\partial w}{\partial y} - \frac{\partial v}{\partial z}\right)\mathbf{i} + \left(\frac{\partial u}{\partial z} - \frac{\partial w}{\partial x}\right)\mathbf{j} + \left(\frac{\partial v}{\partial x} - \frac{\partial u}{\partial y}\right)\mathbf{k} \tag{3.7}$$

and

$$\mathbf{D} \equiv \left(\frac{\partial w}{\partial y} + \frac{\partial v}{\partial z}\right)\mathbf{i} + \left(\frac{\partial u}{\partial z} + \frac{\partial w}{\partial x}\right)\mathbf{j} + \left(\frac{\partial v}{\partial x} + \frac{\partial u}{\partial y}\right)\mathbf{k}. \tag{3.8}$$

Using (3.7) and (3.8), we can rewrite (3.5) as

$$\alpha_o \nabla^2 p' = -\left[\left(\frac{\partial u}{\partial x}\right)^2 + \left(\frac{\partial v}{\partial y}\right)^2 + \left(\frac{\partial w}{\partial z}\right)^2\right] - \frac{1}{2}\left[|\mathbf{D}|^2 - |\boldsymbol{\omega}|^2\right] + \frac{\partial B}{\partial z}, \tag{3.9}$$

and for "well-behaved" fields (i.e., $\nabla^2 p' \propto -p'$),

$$p' \propto \left[\left(\frac{\partial u}{\partial x}\right)^2 + \left(\frac{\partial v}{\partial y}\right)^2 + \left(\frac{\partial w}{\partial z}\right)^2\right] + \frac{1}{2}|\mathbf{D}|^2 - \frac{1}{2}|\boldsymbol{\omega}|^2 - \frac{\partial B}{\partial z}. \tag{3.10}$$

From (3.10) we see that low pressure is associated with rotation (of any sense) by way of the $|\boldsymbol{\omega}|^2$ term. We also see that deformation is associated with high perturbation pressure. We know that hydrostatically, warming in a column leads to pressure falls in the region below the warming. The $\partial B/\partial z$ term partly accounts for such hydrostatic effects. Low (high) pressure perturbations occur below (above) regions of maximum buoyancy (e.g., below and above the region of maximum latent heat release). Lastly, the first term on the righthand side of (3.10), which contains squares of $\partial u/\partial x$, etc., is called the "fluid extension term." From it, high perturbation pressure is associated with both convergence and divergence.

Pressure fluctuations associated with the first three terms on the righthand side of (3.10) sometimes are referred to as "dynamic pressure perturbations," p'_d, while pressure perturbations associated with the fourth term on the righthand side of (3.10) sometimes are referred to as "buoyancy pressure perturbations," p'_b, where

$$p' = p'_d + p'_b. \tag{3.11}$$

Comparison of the decomposition of pressure perturbations in this section with that performed in the previous section [compare (2.1) with (3.11)] reveals that the nonhydrostatic pressure perturbation, p'_{nh}, comprises the dynamic pressure perturbation, p'_d, and *a portion of* the buoyancy pressure perturbation, p'_b. The hydrostatic pressure perturbation, p'_h, comprises the remainder of p'_b.

It is worth noting that the terms on the righthand side of (3.10) should not be viewed as "forcing terms" that *cause* pressure to change. This is because (3.10) is a *diagnostic* equation rather than a *prognostic* equation. It is more precise to say that the terms on the righthand side of (3.10) are *associated with* pressure fluctuations of various sign and magnitude.

4 Bernoulli equations

The assumption that $\nabla^2 p' \propto -p'$ generally is satisfactory for well-behaved fields and away from the ground, but near the ground (a physical boundary), the diagnostic pressure equation we derived in the previous section may not necessarily predict the correct pressure perturbation, owing to a lack of consideration of boundary conditions in making the $\nabla^2 p' \propto -p'$ assumption. Pressure perturbations must be inferred using some other means, and a class of equations called Bernoulli equations often are useful. Bernoulli equations can be formulated if the flow is steady or if it is irrotational (the flow also must be inviscid and in an inertial reference frame, i.e., no friction or Coriolis force). Here we will look at the steady flow case ($\partial/\partial t = 0$).

Let us again write the momentum equation, this time as

$$\frac{\partial \mathbf{v}}{\partial t} + \mathbf{v} \cdot \boldsymbol{\nabla} \mathbf{v} = -\frac{1}{\rho} \boldsymbol{\nabla} p - g\mathbf{k}. \tag{4.1}$$

If the flow is steady, then $\partial \mathbf{v}/\partial t = 0$. Also, we can use vector identities to express $\mathbf{v} \cdot \boldsymbol{\nabla} \mathbf{v}$ as $\boldsymbol{\nabla}|\mathbf{v}|^2/2 + \boldsymbol{\omega} \times \mathbf{v}$. Furthermore, we can write $g\mathbf{k}$ as $\boldsymbol{\nabla} gz$. We thus obtain

$$\boldsymbol{\nabla}\frac{|\mathbf{v}|^2}{2} + \boldsymbol{\omega} \times \mathbf{v} + \frac{1}{\rho}\boldsymbol{\nabla} p + \boldsymbol{\nabla} gz = 0. \tag{4.2}$$

Hereafter we will write $|\mathbf{v}|$ as v. If we take $d\mathbf{r}\cdot$(4.2), where $d\mathbf{r} = (dx\,\mathbf{i} + dy\,\mathbf{j} + dz\,\mathbf{k})$ ($d\mathbf{r}$ is an infinitesimal distance along a streamline), and note that $d\mathbf{r} \cdot \boldsymbol{\omega} \times \mathbf{v} = 0$ because $\boldsymbol{\omega} \times \mathbf{v}$ points normal to a streamline, then (4.2) becomes

$$d\left(\frac{v^2}{2}\right) + \frac{dp}{\rho} + d(gz) = 0, \tag{4.3}$$

and integrating (4.3) gives

$$\frac{v^2}{2} + \int \frac{dp}{\rho} + gz = \text{constant}, \tag{4.4}$$

where the lefthand side of (4.4) is conserved along streamlines (which also are trajectories since $\partial/\partial t = 0$). Equation (4.4) is called a Bernoulli equation. If ρ is constant, then (4.4) can be written as

$$\frac{v^2}{2} + \frac{p}{\rho} + gz = \text{constant}, \tag{4.5}$$

which is another Bernoulli equation.

If the equation of motion is written in terms of the perturbation pressure gradient and buoyancy forces, i.e.,

$$\frac{\partial \mathbf{v}}{\partial t} + \mathbf{v} \cdot \nabla \mathbf{v} = -\frac{1}{\rho}\nabla p' + B\mathbf{k}, \tag{4.6}$$

then instead of (4.4) we obtain a Bernoulli equation of the form

$$\frac{v^2}{2} + \int \frac{dp'}{\rho} - \int B\, dz = \text{constant}. \tag{4.7}$$

Notice in (4.7) that if we neglect pressure perturbations and integrate B from the level of free convection (LFC) to the equilibrium level (EL), we obtain the parcel theory estimate of the maximum velocity in an updraft:

$$\frac{v^2}{2} = \int_{LFC}^{EL} B\, dz \tag{4.8}$$

$$v = \sqrt{2\,\text{CAPE}}, \tag{4.9}$$

where CAPE is the convective available potential energy.

5 Further reading

Das, P., 1979: A non-Archimedian approach to the equations of convection dynamics. *J. Atmos. Sci.*, **36**, 2183–2190.

Davies-Jones, R., 2003: An expression for effective buoyancy in surroundings with horizontal density gradients. *J. Atmos. Sci.*, **60**, 2922–2925.

Doswell, C. A., and P. M. Markowski, 2004: Is buoyancy a relative quantity? *Mon. Wea. Rev.*, **132**, 853–863.

List, R. and E. P. Lozowski, 1970: Pressure perturbations and buoyancy in convective clouds. *J. Atmos. Sci.*, **27**, 168–170.

Convective Storm Initiation and Organization

Paul Markowski
Department of Meteorology, Penn State University, University Park, Pennsylvania, USA

Abstract. The conditions necessary for the development of deep moist convection are discussed. This material is followed by an overview of the ways in which convective storms may be organized, with emphasis on the environmental characteristics favoring the various organizational modes.

1 Introduction

Convective storms can be organized in a wide variety of ways (Fig. 1). Some occur as discrete cells, whereas others occur within large complexes or lines. The organization of convective storms, to considerable degree, influences the threats they pose. The primary threat of many storms is heavy rain, whereas in others, hail or high winds may be the primary threats. In other classes of storms, tornadoes may pose a serious threat.

If one is to anticipate the threat of severe weather, one must correctly anticipate whether convective storms are likely to be initiated in a given region, and secondly, one must determine the range of likely storm types. Perhaps ironically, the former tends to be more difficult than the latter. Determining where and when convective storms are likely to be initiated is likely a complex function of vertical motions ranging from the scale of thermals to the synoptic scale, temperature and moisture inhomogeneities, and the mean stratification largely resulting from synoptic-scale processes. Determining the mode of convective organization, once convection has been initiated, depends largely on the amount of vertical wind shear present in environment of the convection.

2 Convection initiation

The process of initiating a deep convective cloud has been referred to as "convection initiation." The process involves parcels of air reaching a level of free convection (LFC), and then achieving and maintaining positive buoyancy over a significant upward vertical excursion.

The presence of an LFC and convective available potential energy (CAPE) requires relatively large lower to middle tropospheric lapse rates (the average lapse rate must be larger than the moist adiabatic lapse rate) and low-level moisture. The convection initiation "problem" stems from the fact that the presence of CAPE is not sufficient for convection initiation. Air typically requires some forced ascent in order to reach its LFC, owing to the presence of at least some convective inhibition (CIN) on most environmental soundings. There are many mechanisms known to initiate thunderstorms, such as lifting by convergent boundaries such as fronts, drylines, outflow boundaries, sea breezes, and land breezes (especially intersections of horizontal convective rolls with one of these boundaries), ascent forced by flow over topography, circulations forced by differential heating (cloudy-clear air boundaries, heating of sloped terrain, horizontal sensible heat flux variations), and forced lifting by gravity waves.

Synoptic-scale dynamics often prime the mesoscale environment for convection initiation (synoptic-scale dynamics also can inhibit convection initiation). The presence of large-scale upward motion, which may be anticipated reasonably well from model guidance and from the diagnosis of quasigeostrophic forcings, leads to the reduction of CIN. On the other hand, if there is synoptically-driven, large-scale *descent*, then CIN tends to be augmented and convection initiation becomes less likely. Thus, in forecasting convection initiation—a distinctly mesoscale phenomenon—it often is important to identify *synoptic-scale* features such as jet streaks and regions of strong thermal and differential vorticity advection.

Several complexities make the task of real-time prediction of the development of convective storms rather arduous. A complete analysis of these complexities is beyond the scope of this lecture, and likely awaits future research anyway. In short, some of the complexities arise from the fact that mixing occurs en route to the saturation level and LFC along air parcel trajectories. Thus, convection initiation, as mentioned in a previous lecture, is not as simple as reaching the so-called "convective temperature" or having CIN$\rightarrow$0 J kg^{-1}. Furthermore, it is not possible to know the three-dimensional structure of vertical velocity fields along mesoscale boundaries, nor is it even easy to grasp the extent of mesoscale temperature and moisture heterogeneity on the mesoscale (neither in real-time nor in ex post facto studies). Perhaps the most practical forecasting strategy for predicting convection initiation (assuming CAPE is present) is to be vigilant of regions of persistent low-level convergence where CIN values are small (or inferred to be small, based on comparisons of observed surface conditions with morning sounding data or model sounding data).

3 Convection organization

Once one has determined if and where deep moist convection will occur, the next logical step is to forecast the type of storms that are most likely. Compared to forecasting whether or not storms will form in the first place, predicting storm type might be considered by some to be relatively easy. Vertical shear tends to promote storm organization and longevity, although excessive vertical shear can be detrimental to weak updrafts in environments of marginal instability. Although there are many measures of vertical wind shear or its associated horizontal vorticity, one of the least volatile parameters for evaluating vertical shear and predicting storm type is simply the magnitude of the 0–6 km shear vector (the 1000–500 mb shear vector length is a suitable proxy). This parameter, unlike many others (e.g., hodograph length, storm-relative helicity), is insensitive to kinks and inflections in the hodograph, which are fractal. Small-scale hodograph details quite possibly are important for some sorts of predictions (e.g., whether or not a tornado might develop), but for anticipating storm type, small-scale hodograph details do not appear to be as critical. The 0–6 km shear vector magnitude is easy to compute, since it requires no integrations or calculations of mean winds.

3.1 Ordinary convection

"Ordinary" convection is the term for the variety of deep moist convection that is single-celled and has a lifetime on the order of 45 minutes. It tends to occur near and shortly after the time of maximum daytime heating and quickly dissipates after sunset. Ordinary convection only occasionally produces hail or wind gusts that could be characterized as severe. When severe weather is produced, it generally is of the "pulse" variety—short-lived, usually marginal (e.g., a brief wind gust >25 m s^{-1}), and difficult to issue warnings for. Updraft speeds in ordinary convective cells usually range from 5–20 m s^{-1}.

Although environmental CAPE may range from just a few hundred J kg^{-1} to >2000 J kg^{-1}, ordinary convection occurs in the absence of appreciable wind shear (i.e., the magnitude of the vector wind difference between the surface and 6 km typically is ≤ 10 m s^{-1}. The lifetime of an ordinary convective cell, τ, can be approximated as the time it takes air to ascend from the surface to the equilibrium level (approximately H, the scale height of the atmosphere) divided by the average updraft speed (w_o), plus the time it takes for the precipitation produced by the ascent to fall to the ground, which can be approximated as H/v_t, where v_t is the mean terminal fallspeed of the precipitation:

$$\tau \approx \frac{H}{w_o} + \frac{H}{v_t}. \tag{3.1}$$

For $H \simeq 10$ km, $w_o \simeq 5$–10 m s^{-1}, and $v_t \simeq 5$–10 m s^{-1}, the lifetime is roughly 30–60 minutes. The lifetimes of individual cells are limited to roughly an hour or less owing to the spread of evaporatively cooled air near the surface. The outflow faces little resistance from relative inflow because relative inflow is weak in weak-shear environments. Furthermore, the lack of significant vertical wind shear leads to precipitation that falls back through the updraft rather than being advected downstream, which occurs when substantial anvil-level storm-relative winds are present (and they can only be present in environments containing significant wind shear values).

The lifecycle of an ordinary convective cell begins with the "towering cumulus stage." In this stage, only an updraft exists. The towering cumulus stage is followed by the "mature stage," which commences with the production of precipitation particles large enough to fall through the updraft. At this time, an anvil also begins to form. The precipitation formed by the rising air currents within the updraft falls *through* the updraft rather than being deposited away from the updraft, and the associated "water loading" reduces updraft buoyancy. The falling precipitation and subsequent evaporation of precipitation induce a downdraft, which spreads laterally upon reaching the surface. The leading edge of the downdraft marks the gust front.

Eventually the downdraft completely dominates the cell, at which time the "dissipating stage" begins. During this stage, the rain-cooled air spreads far from the updraft; the updraft is cut off from potentially buoyant inflow and cannot be maintained. New cells may or may not be initiated by the gust front, depending on the environmental CIN and the vertical motion associated with the gust front. Eventually the old convective cloud is reduced to an "orphan anvil" composed entirely of ice crystals. The anvil slowly sublimates into oblivion (this may take many hours).

If severe weather occurs, it usually occurs near the transition from the mature to dissipating stages, or about the time when the maximum fallout of precipitation occurs. Severe wind gusts are most likely when precipitation falls through relatively dry subcloud layers, where evaporative cooling and generation of negative buoyancy is substantial.

As the magnitude of the vertical wind shear increases, the precipitation-driven downdraft interferes less with the updraft, and the forward speed of the gust front relative to the updraft decreases—the result is an increase in the intensity and organization of the convection. However, the *formation* of storms may be inhibited if the vertical wind shear becomes excessively large.

3.2 Multicellular convection

Multicellular convection probably is the most common type of convection in midlatitudes. In contrast to "ordinary" convection, multicellular convection consists of numerous updrafts in various stages of maturity. The entire system of updrafts may last for hours and produce damaging straight-line winds and hail up to the size of golf balls. Multicellular convection can be organized as clusters of updrafts or as lines that may be broken or nearly continuous. A detailed survey of all of the many forms of multicellular convection (e.g., squall lines, mesoscale convective complexes, bow echoes) and their associated severe weather threats is beyond the scope of this lecture.

Multicellular convection occurs in environments containing a moderate amount of vertical wind shear—the 0-6 km shear vector magnitude tends to lie in the 10–20 m s^{-1} range. CAPE values may vary considerably, from 500–5000 J kg^{-1}. The presence of a moderate amount of wind shear promotes organization within the convective system, which allows the system to persist as a whole for relatively long periods of time. The defining radar characteristic of multicell storms is a series of echoes in various stages of evolution. Individual echoes, which are associated with discrete updrafts, tend to move at the mean wind speed over the updraft depth. Where the lifting by the gust front is sufficient to overcome the ambient CIN, new cells are triggered, thereby enabling the survival of a larger-scale convective system. New cells tend to be forced on the downshear flank of the gust front, because lifting by a gust front (which behaves like a density current) is enhanced where the ambient shear over approximately the depth of the outflow is in line with the density gradient associated with the outflow. Although new cell development is favored on the downshear flank by the low-level shear, the precise location where new cells are successively initiated by the gust front—and overall system propagation—also may be affected by environmental heterogeneity, especially in CIN. Midlevel vertical wind shear also can be important because it can lead to updraft-relative flow at low levels, which has the effect of restraining the outflow (i.e., slowing the gust front's motion) with respect to the updraft, thereby slowing the rate at which the updraft is undercut by outflow. The maintenance of multicellular convection, particularly long-lived squall lines, remains an intensely studied subject today, with much of the current interest being in the dynamical importance of middle and upper-tropospheric wind shear, the vertical distribution of wind shear and system-relative winds, and the importance of the components of the wind shear and system-relative winds that are parallel to lines of updrafts.

3.3 Supercellular convection

When the vertical wind shear becomes larger than approximately 20 m s^{-1} over the lowest half of the troposphere, isolated convection develops into what have been coined "supercells." Supercells contain organized, storm-scale rotation and are greatly influenced by dynamic pressure gradients. They tend to be the most severe variety of convective storm, responsible for virtually all strong tornadoes and hail diameters in excess of 5 cm. A more thorough summary of supercells and their associated dynamics is reserved for another lecture.

4 Final remarks

It is worth noting that most of the present understanding summarized above is based on idealized numerical simulations in which the environments were horizontally homogeneous and time invariant, and storms were initiated with isolated warm bubbles. A growing body of observational evidence and recent idealized numerical simulations suggest that spatial inhomogeneities in the environment can lead to profound changes in convective storm morphology and behavior. Furthermore, synoptic and mesoscale conditions are rarely, if ever, unchanging in time. We do not have a good grasp of what happens when convection initiated in weak shear finds itself in much stronger shear hours later, nor do we have a mature understanding of the sensitivity of convection organization to the processes responsible for convection initiation.

5 Further reading

Doswell, C. A., 2001: Severe convective storms—An overview. *Severe Local Storms, Meteor. Monogr.*, No. 50, 1–26.

Fritsch, J. M., and G. S. Forbes, 2001: Mesoscale convective systems. *Severe Local Storms, Meteor. Monogr.*, No. 50, 323–358.

Weisman, M. L., and J. B. Klemp, 1982: The dependence of numerically simulated convective storms on vertical wind shear and buoyancy. *Mon. Wea. Rev.*, **110**, 504–520.

Weisman, M. L., and J. B. Klemp, 1984: The structure and classification of numerically simulated convective storms in directionally varying wind shears. *Mon. Wea. Rev.*, **112**, 2479–2498.

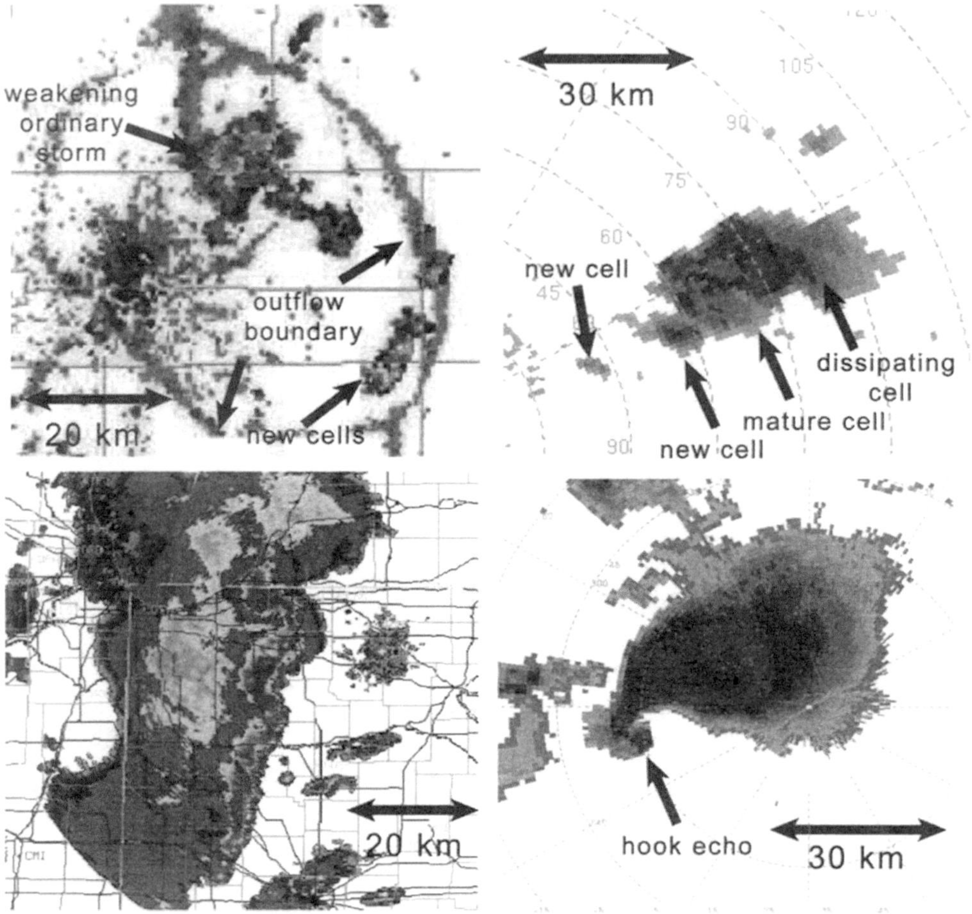

Figure 1. Examples of ordinary convection (top left), multicellular convection (top right), a squall line (lower left), and a supercell thunderstorm (lower right), as viewed in radar reflectivity data.

Supercell Thunderstorms

Paul Markowski

Department of Meteorology, Penn State University, University Park, Pennsylvania, USA

Abstract. The environments and characteristics of supercell thunderstorms are reviewed, followed by a theoretical discussion of the dynamical origins of supercell rotation and propagation. Supercell thunderstorm environments are characterized by large vertical wind shear. The large wind shear promotes longevity, organization, and severity for several reasons: (1) large storm-relative winds associated with the wind shear minimize the degree to which precipitation interferes with the updraft; (2) large horizontal vorticity associated with the wind shear is tilted into the vertical to produce a midlevel mesocyclone; (3) the rotation within the updraft and the interaction of the updraft with the large environmental wind shear give rise to strong dynamic pressure gradients which can feed back to the intensity and organization of the supercell.

1 Basic characteristics and environments

The term "supercell" was coined in the middle 1970s[1], although supercells were first identified as a special class of convective storm in the early 1960s. Supercells are the least frequent type of storm, but they produce a disproportionately large amount of severe weather, including large hail, damaging wind, intense lightning, and tornadoes. Almost all reports of hail having a diameter of 5 cm or larger are associated with supercell storms. Virtually all strong and violent tornadoes also are produced by supercells. Supercells are associated with some of the highest lightning flash rates ever recorded, with rates exceeding 200 flashes per minute possible.

We probably know more about supercell dynamics than the dynamics of most other precipitating convection, with the possible exception of squall lines. This probably is attributable, in part, to the steadiness of supercell storms. Supercells are frequently long-lived (1–4 h lifetimes are common), in fact, this is probably one reason why they were originally called "supercells." The consensus within the scientific community now seems to favor abandoning a longevity criterion for classification of a storm as a supercell in favor of a dynamical criterion which requires the presence of a persistent, deep mesocyclone. A mesocyclone is a region of vertical vorticity with a characteristic width of 3–8 km and magnitude of $O(10^{-2})$ s^{-1}. By "persistent" we mean long enough for an air parcel to pass through the entire updraft (~20 min), and by "deep" we mean extending over at least half the depth of the updraft. Curiously, the motion of supercells tends to deviate

[1]K. Browning is usually given credit for this nomenclature.

significantly from the mean wind, with cyclonically (anticyclonically) rotating supercells having a propagation component to the right (left) of the mean wind shear.

Supercells occur in environments containing large vertical wind shear. In other words, long hodographs typify supercell environments. Hodographs often also are curved, although hodograph curvature is not necessary for supercell storms (Fig. 1). In contrast to many squall line environments having large low-level wind shear beneath weak middle to upper tropospheric wind shear, the vertical wind shear present in supercell environments tends to be relatively evenly distributed with respect to the vertical. The vertical wind shear profile is associated with significant storm-relative winds at all levels. Convective available potential energy (CAPE) tends to be significant (>1000 J kg^{-1}) in supercell environments, although extreme amounts of CAPE are not necessary.

The presence of strong deep-layer shear and strong storm-relative winds leads to the deposition of precipitation away from the updraft. Furthermore, strong low-level storm-relative winds restrain the surface gust front and prevent it from undercutting the updraft. Dynamically-induced vertical pressure gradients arising from the interaction of an updraft with vertical wind shear also can promote upward motion, especially on storm flanks. Thus, the presence of strong deep-layer vertical shear can promote long-lived updrafts. Moreover, the horizontal vorticity associated with the large vertical wind shear in supercell environments is the source of vertical vorticity for the midlevel mesocyclone by way of the tilting term in the vorticity equation.

2 Structure

Supercells are characterized by a single, dominant updraft, which may have a quasi-steady appearance, depending on the temporal frequency of observations. The updraft cores may contain intense vertical velocities, sometimes as large as 60 m s^{-1}. The supercell updraft also contains rotation, referred to as a mesocyclone. The updraft and mesocyclone region often are visually stunning (Fig. 2), and in Doppler radar velocity data, the mesocyclone is associated with a couplet of inbound and outbound radial velocities. A wall cloud may be present at the base of the updraft, where humid, rain-cooled air originating from the forward-flank precipitation region is drawn into the updraft.

The updraft of a supercell typically is associated with a reflectivity minimum in radar data, called the bounded weak-echo region (BWER). The BWER is assumed to signify a region of updraft too strong to allow the descent of hydrometeors and/or a region in which the Lagrangian time scales are insufficient for precipitation formation. At low levels, the BWER may not be "bounded," in which case a WER, or weak-echo region exists.

The downward extension of the rear side of the "echo overhang," which caps the BWER, forms a pendant-shaped "hook echo," as viewed by radar (Fig. 3). The hook echo probably is the best-recognized reflectivity feature associated with a supercell storm. It is not yet clear to what degree this echo protrusion is due to lateral extrusion of precipitation from the main echo region, versus the progressive downward development of precipitation to the rear of the WER.

Supercells contain two main downdraft regions. The first is associated with the hook echo region to the rear of the storm, and this downdraft is referred to as the "rear-flank downdraft," or RFD (Fig. 4). It has long been surmised that the RFD forms when dry mid- and upper-level winds impinge upon the backside of the updraft, leading to evaporative chilling and negative buoyancy, and ensuing downward accelerations. However, downward-directed dynamic vertical pressure gradients on the upshear flank of the storm probably also play a role in driving the RFD, and the extent to which RFDs are driven thermodynamically versus dynamically is not yet fully understood. If low-θ_e air at midlevels is entrained into the RFD, low equivalent potential temperature (θ_e) values may be observed at the surface (θ_e deficits of 10–20 K may be observed, depending on θ_e values aloft and the degree of entrainment). If the RFD is driven predominantly by precipitation loading and/or dynamic effects, then θ_e deficits within the RFD may not be as substantial (e.g., 2–5 K). Temperature deficits in the RFD have a wide range of values, depending on the extent of entrainment of potentially cold air aloft and the amount of latent cooling within the RFD (temperature deficits may range from near zero to 10 K).

The upper-level storm-relative winds deposit hydrometeors on the forward flank of the updraft. Evaporation of rain and the melting and sublimation of hail lead to the development of negative buoyancy and a downdraft in this forward-flank region. This downdraft appropriately is referred to as the "forward-flank downdraft," or FFD (Fig. 4). In dry ambient environments, greater cooling potential exists, so that FFDs tend to be prominent. In moist ambient environments, the evaporative cooling potential is limited and an FFD may not be well-defined. Temperature and θ_e deficits within the FFD may have a wide range of values, e.g., 2–10 K, depending on the large-scale low-level relative humidity and degree to which air that is latently chilled aloft reaches the ground. The FFD and RFD collectively produce a surface gust front structure that kinematically resembles the frontal structure associated with a midlatitude extratropical cyclone (Fig. 4).

3 Origins of rotation

3.1 Linear theory

The origin of midlevel rotation in supercells is relatively well understood. We begin with the vertical vorticity equation written as

$$\begin{aligned}\frac{\partial \zeta}{\partial t} &= -\mathbf{v}\cdot\nabla\zeta + \boldsymbol{\omega}\cdot\nabla w \qquad (3.1)\\ &= -u\frac{\partial \zeta}{\partial x} - v\frac{\partial \zeta}{\partial y} - w\frac{\partial \zeta}{\partial z} + \xi\frac{\partial w}{\partial x} + \eta\frac{\partial w}{\partial y} + \zeta\frac{\partial w}{\partial z}\\ &= -u\frac{\partial \zeta}{\partial x} - v\frac{\partial \zeta}{\partial y} - w\frac{\partial \zeta}{\partial z} + \left(\frac{\partial w}{\partial y} - \frac{\partial v}{\partial z}\right)\frac{\partial w}{\partial x} + \left(\frac{\partial u}{\partial z} - \frac{\partial w}{\partial x}\right)\frac{\partial w}{\partial y} + \left(\frac{\partial v}{\partial x} - \frac{\partial u}{\partial y}\right)\frac{\partial w}{\partial z}\end{aligned}$$

where the velocity vector is $\mathbf{v} = (u, v, w)$, the vorticity vector is $\boldsymbol{\omega} = (\xi, \eta, \zeta)$, and the Coriolis force and baroclinic effects have been neglected (both can be shown to be small compared to the advection and tilting/stretching terms). We can mathematically explain mesocyclogenesis by way of a simple, linear vertical vorticity equation, obtained by linearizing (3.1) about a mean flow containing vertical wind shear. In other words, let $u = \bar{u}(z) + u'$, $v = \bar{v}(z) + v'$, $w = w'$, and $\zeta = \zeta'$,and neglecting the products of perturbations yields

$$\frac{\partial \zeta'}{\partial t} = -\bar{u}\frac{\partial \zeta'}{\partial x} - \bar{v}\frac{\partial \zeta'}{\partial y} + \frac{\partial \bar{u}}{\partial z}\frac{\partial w'}{\partial y} - \frac{\partial \bar{v}}{\partial z}\frac{\partial w'}{\partial x}; \tag{3.2}$$

thus,

$$\frac{\partial \zeta'}{\partial t} = -\bar{\mathbf{v}} \cdot \boldsymbol{\nabla}_h \zeta' + \mathbf{S} \times \boldsymbol{\nabla}_h w' \cdot \mathbf{k}, \tag{3.3}$$

where the h subscript indicates horizontal and the vertical wind shear is $\mathbf{S} = \partial \bar{\mathbf{v}}_h/\partial z$.

It is easier to interpret (3.3) if it is rewritten so that $\partial \zeta'/\partial t$ is considered in the (moving) reference frame of the updraft. If the updraft moves with velocity $\mathbf{c}$ (note that $\mathbf{c}$ is a constant), subtraction of $\mathbf{c}$ from $\bar{\mathbf{v}}$ wherever $\bar{\mathbf{v}}$ appears yields

$$\frac{\partial \zeta'}{\partial t} = -(\bar{\mathbf{v}} - \mathbf{c}) \cdot \boldsymbol{\nabla}_h \zeta' + \mathbf{S} \times \boldsymbol{\nabla}_h w' \cdot \mathbf{k}, \tag{3.4}$$

where $\bar{\mathbf{v}} - \mathbf{c}$ is the *storm-relative* wind. The second term on the righthand side of (3.4) is responsible for generating ζ' by way of tilting of horizontal vorticity associated with the mean vertical wind shear by vertical velocity gradients. The first term on the righthand side of (3.4) represents advection of ζ' by the storm-relative wind. This term shifts the ζ' field horizontally within the updraft once ζ' has been generated (note that only the $\mathbf{S} \times \boldsymbol{\nabla}_h w' \cdot \mathbf{k}$ term can *generate* ζ', since the $-(\bar{\mathbf{v}} - \mathbf{c}) \cdot \boldsymbol{\nabla}_h \zeta'$ term is zero when ζ' is everywhere zero). As an updraft intensifies, vorticity that has been tilted into the vertical also is amplified substantially by stretching ($\zeta' \, \partial w'/\partial z$), which is a nonlinear effect, neglected in (3.4).

A more detailed qualitative examination of (3.4) reveals that the $\mathbf{S} \times \boldsymbol{\nabla}_h w' \cdot \mathbf{k}$ term produces a vorticity couplet that straddles the location of w_{max} (the maximum updraft), with $\zeta' > 0$ ($\zeta' < 0$) on the right (left) side of w_{max}, if looking downshear (Fig. 5). (Use the righthand rule to convince yourself that the above relationship is correct—the vorticity couplet produced by tilting is aligned orthogonally with respect to the shear direction.)

Now consider more carefully the $-(\bar{\mathbf{v}} - \mathbf{c}) \cdot \boldsymbol{\nabla}_h \zeta'$ term in (3.4). Note that $\boldsymbol{\nabla}_h \zeta'$ points in the *opposite* direction of $\boldsymbol{\omega}_h$, i.e., 90° to the right of $\mathbf{S}$ (Fig. 6). When $\bar{\mathbf{v}} - \mathbf{c}$, the storm-relative wind, points in the same direction as $\boldsymbol{\omega}_h$, we have *streamwise vorticity* by definition, and $-(\bar{\mathbf{v}} - \mathbf{c}) \cdot \boldsymbol{\nabla}_h \zeta' > 0$. Therefore, $\partial \zeta'/\partial t > 0$ at the location of w_{max}, which shifts the cyclonic vortex ($\zeta' > 0$) of the ζ' couplet toward the location of w_{max} (i.e., w_{max} is collocated with cyclonic vorticity). On the other hand, when $\bar{\mathbf{v}} - \mathbf{c}$ points in the direction normal to $\boldsymbol{\omega}_h$, we have *crosswise vorticity* by definition, and $-(\bar{\mathbf{v}} - \mathbf{c}) \cdot \boldsymbol{\nabla}_h \zeta' = 0$. Therefore, $\partial \zeta'/\partial t = 0$ at the location of w_{max}, and thus, the vorticity couplet arising from horizontal vorticity tilting is not shifted laterally with respect to the location of w_{max} (i.e., the vorticity couplet straddles w_{max})(Fig. 7).

Summarizing the generation of rotation aloft within a thunderstorm updraft, initially a vorticity couplet develops due to tilting of horizontal vorticity. The couplet initially straddles the updraft maximum, because the $-(\bar{\mathbf{v}} - \mathbf{c}) \cdot \boldsymbol{\nabla}_h \zeta'$ term in (3.4) is zero initially.

Later in the evolution of the storm (e.g., after ~30–60 minutes), depending on the degree to which the horizontal vorticity has a streamwise component, the ζ' field shifts laterally relative to the updraft maximum, so that the rotation and updraft may become more in phase. The relative magnitudes of the streamwise versus crosswise vorticity components is sensitive to storm motion. For example, once deviant storm motion commences, the horizontal vorticity typically becomes more streamwise with respect to the deviant motion. We'll soon find that the rotation which develops within supercells leads to pressure perturbations and gradients of pressure perturbations, which in turn affect the motion of updrafts themselves. In other words, updraft rotation and propagation are intimately related to each other, and it is the nature of this relationship that we'll examine shortly.

3.2 Potential vorticity conservation

Another easy was to conceptualize the tilting process is by way of Ertel's potential vorticity conservation:

$$\frac{d}{dt}\left(\frac{\boldsymbol{\omega}\cdot\nabla\theta_e}{\rho}\right)=0, \tag{3.5}$$

where ρ is the air density and θ_e is an appropriate conservative variable for dry and moist adiabatic motions; however, θ_e is not a state variable for dry adiabatic motions, therefore (3.5) is only approximately correct in subsaturated regions.[2]

In most environments, θ_e varies much more in the vertical than in the horizontal direction (e.g., θ_e may decrease from ~350 K near the ground to ~330 K just a few kilometers above the ground, whereas θ_e typically varies by less than 10 K over horizontal distances of 100–500 km, at least prior to storm formation); thus, $\nabla\theta_e$ is directed approximately vertically (pointing downward into the ground). In supercell environments, the horizontal vorticity components (ξ and/or η) typically are 1–2 orders of magnitude larger than the pre-storm ζ values (~10^{-2} s^{-1} versus ~10^{-3}–10^{-4} s^{-1}); thus, $\boldsymbol{\omega}$ tends to be oriented quasi-horizontally. Therefore, $\boldsymbol{\omega}\cdot\nabla\theta_e \approx 0$ in the pre-storm environment, and Ertel's potential vorticity conservation tells us that $\boldsymbol{\omega}\cdot\nabla\theta_e \approx 0$ for all time. Since $\boldsymbol{\omega}\cdot\nabla\theta_e \approx 0$ means that vortex lines lie within θ_e surfaces, then vortex lines lie within θ_e surfaces for all time (in the absence of mixing). In a thunderstorm, θ_e surfaces erupt upward (downward) within the updraft (downdraft); thus, vertical vorticity is produced by tilting in regions where θ_e surfaces slope upward (and downward). Whereas (3.3) applies only to linear flows, (3.5) applies to fully nonlinear flows (e.g., those that contain significant vorticity stretching).

3.3 Rotation at low levels

Numerical simulations conducted in the early to middle 1980s found that the baroclinic zones produced on the forward flanks of supercells may generate significant hor-

[2]To be exact, the righthand side of (3.5) should be $-\alpha J(B,\theta_e)$, where $\alpha = 1/\rho$, B is buoyancy, and J is the Jacobean operator. When air is saturated, $B = B(\theta_e, z)$ and $-\alpha J(B,\theta_e) = 0$. When air is unsaturated (e.g., for dry adiabatic motions), $-\alpha J(B,\theta_e)$ has been shown to remain relatively small, so that (3.5) is reasonably accurate for both dry and moist adiabatic motions.

izontal vorticity. This horizontal vorticity may add to the horizontal vorticity already present on the larger scale that is associated with the mean vertical wind shear. In other words, the forward-flank baroclinic zone may provide "extra" or augmented horizontal vorticity that is available for tilting by the updraft. The increased vertical vorticity production owing to increased tilting (and subsequent stretching) may lead to the development of a mesocyclone at low levels (say, 1–3 km above ground level). Without a source of horizontal vorticity augmentation (such as the forward flank baroclinic zone), it has been found that significant vertical vorticity is confined to midlevels (say, 3–7 km) within the storm.

4 Supercell propagation

4.1 Dynamic pressure perturbations

To investigate the dynamics that give rise to supercell propagation, we begin with a look at a diagnostic pressure equation. If we take the divergence ($\boldsymbol{\nabla}\cdot$) of the momentum equations, neglect the Coriolis force, and make the Boussineq approximation in order to facilitate qualitative interpretation, it can be shown that

$$\alpha_o \nabla^2 p' = -\left[\left(\frac{\partial u}{\partial x}\right)^2 + \left(\frac{\partial v}{\partial y}\right)^2 + \left(\frac{\partial w}{\partial z}\right)^2\right] - 2\left(\frac{\partial v}{\partial x}\frac{\partial u}{\partial y} + \frac{\partial w}{\partial x}\frac{\partial u}{\partial z} + \frac{\partial w}{\partial y}\frac{\partial v}{\partial z}\right) + \frac{\partial B}{\partial z} \tag{4.1}$$

where α_o is the specific volume, B is the buoyancy, and p' is the deviation of pressure from a horizontally homogeneous, hydrostatic base state. Let us decompose the wind field into a mean flow with shear and perturbation parts; i.e., let $u = \bar{u} + u'$, $v = \bar{v} + v'$, and $w = w'$. Then (4.1) becomes (we will retain the products of perturbations here, for the nonlinear terms will be shown to be important later on)

$$\begin{aligned} \alpha_o \nabla^2 p' &= -\left[\left(\frac{\partial u'}{\partial x}\right)^2 + \left(\frac{\partial v'}{\partial y}\right)^2 + \left(\frac{\partial w'}{\partial z}\right)^2\right] - 2\left(\frac{\partial v'}{\partial x}\frac{\partial u'}{\partial y} + \frac{\partial w'}{\partial x}\frac{\partial u'}{\partial z} + \frac{\partial w'}{\partial y}\frac{\partial v'}{\partial z}\right) \\ &\quad -2\left(\frac{\partial w'}{\partial x}\frac{\partial \bar{u}}{\partial z} + \frac{\partial w'}{\partial y}\frac{\partial \bar{v}}{\partial z}\right) + \frac{\partial B}{\partial z}. \end{aligned} \tag{4.2}$$

On the righthand side, all of the terms except for the $\partial B/\partial z$ terms are associated with what are referred to as "dynamic pressure perturbations." The $\partial B/\partial z$ term is associated with pressure perturbations due to buoyancy. On the righthand side, the first set of terms in [ ] brackets is referred to as the "fluid extension terms;" the second group of terms in ( ) brackets is referred to as the "nonlinear dynamic pressure terms;" the third group of terms in ( ) brackets is referred to as the "linear dynamic pressure terms."

We can rewrite the linear terms as

$$-2\left(\frac{\partial w'}{\partial x}\frac{\partial \bar{u}}{\partial z} + \frac{\partial w'}{\partial y}\frac{\partial \bar{v}}{\partial z}\right) = -2\frac{\partial \bar{\mathbf{v}}_h}{\partial z} \cdot \boldsymbol{\nabla}_h w'. \tag{4.3}$$

We also can simplify the nonlinear terms. Let us suppose that we are interested only in the region within the updraft itself, such that there is no deformation or horizontal vorticity. If there is no deformation, then

$$\begin{aligned} \frac{\partial v'}{\partial x} + \frac{\partial u'}{\partial y} &= 0 \\ \frac{\partial v'}{\partial x} &= -\frac{\partial u'}{\partial y} \end{aligned} \tag{4.4}$$

and

$$\begin{aligned} \frac{\partial w'}{\partial y} + \frac{\partial v'}{\partial z} &= 0 \\ \frac{\partial w'}{\partial y} &= -\frac{\partial v'}{\partial z} \end{aligned} \tag{4.5}$$

and

$$\begin{aligned} \frac{\partial w'}{\partial x} + \frac{\partial u'}{\partial z} &= 0 \\ \frac{\partial w'}{\partial x} &= -\frac{\partial u'}{\partial z}. \end{aligned} \tag{4.6}$$

If there is no horizontal vorticity, then

$$\begin{aligned} \frac{\partial w'}{\partial y} - \frac{\partial v'}{\partial z} &= 0 \\ \frac{\partial w'}{\partial y} &= \frac{\partial v'}{\partial z} \end{aligned} \tag{4.7}$$

and

$$\begin{aligned} \frac{\partial u'}{\partial z} - \frac{\partial w'}{\partial x} &= 0 \\ \frac{\partial u'}{\partial z} &= \frac{\partial w'}{\partial x}. \end{aligned} \tag{4.8}$$

The only way that both (4.5) and (4.7) can be true and both (4.6) and (4.8) can be true is if

$$\frac{\partial u'}{\partial z} = \frac{\partial v'}{\partial z} = \frac{\partial w'}{\partial x} = \frac{\partial w'}{\partial y} = 0. \tag{4.9}$$

The nonlinear term $-2\frac{\partial v'}{\partial x}\frac{\partial u'}{\partial y}$ remains. Using (4.4), we perform the following steps:

$$\begin{aligned}
-2\frac{\partial v'}{\partial x}\frac{\partial u'}{\partial y} &= -\frac{1}{2}\left(4\frac{\partial v'}{\partial x}\frac{\partial u'}{\partial y}\right) \\
&= -\frac{1}{2}\left(\frac{\partial v'}{\partial x}\frac{\partial u'}{\partial y} + 2\frac{\partial v'}{\partial x}\frac{\partial u'}{\partial y} + \frac{\partial v'}{\partial x}\frac{\partial u'}{\partial y}\right) \\
&= -\frac{1}{2}\left(-\frac{\partial v'}{\partial x}\frac{\partial v'}{\partial x} + 2\frac{\partial v'}{\partial x}\frac{\partial u'}{\partial y} - \frac{\partial u'}{\partial y}\frac{\partial u'}{\partial y}\right) \\
&= \frac{1}{2}\left(\frac{\partial v'}{\partial x}\frac{\partial v'}{\partial x} - 2\frac{\partial v'}{\partial x}\frac{\partial u'}{\partial y} + \frac{\partial u'}{\partial y}\frac{\partial u'}{\partial y}\right) \\
&= \frac{1}{2}\left(\frac{\partial v'}{\partial x} - \frac{\partial u'}{\partial y}\right)^2 \\
&= \frac{1}{2}\zeta'^2
\end{aligned} \tag{4.10}$$

Using (4.3), (4.9), and (4.10), we can write (4.2) as

$$\alpha_o\nabla^2 p' = -\left[\left(\frac{\partial u'}{\partial x}\right)^2 + \left(\frac{\partial v'}{\partial y}\right)^2 + \left(\frac{\partial w'}{\partial z}\right)^2\right] + \frac{1}{2}\zeta'^2 - 2\frac{\partial \bar{\mathbf{v}}_h}{\partial z}\cdot\nabla_h w' + \frac{\partial B}{\partial z}. \tag{4.11}$$

We will assume that the pressure field is "well-behaved;" thus, $\nabla^2 p' \propto -p'$ and

$$p' \propto \left[\left(\frac{\partial u'}{\partial x}\right)^2 + \left(\frac{\partial v'}{\partial y}\right)^2 + \left(\frac{\partial w'}{\partial z}\right)^2\right] - \frac{1}{2}\zeta'^2 + 2\frac{\partial \bar{\mathbf{v}}_h}{\partial z}\cdot\nabla_h w' - \frac{\partial B}{\partial z}. \tag{4.12}$$

The terms on the righthand side of (4.12) are the fluid extension terms, the nonlinear dynamic term, linear dynamic term, and buoyancy term. Although they may appear to be "forcings" for p', (4.12) is a *diagnostic* equation; i.e., the terms on the righthand side are *associated with* pressure fluctuations (they do not *cause* the pressure fluctuations). All of the terms on the righthand side of (4.12) may contribute equally to p', but the nonlinear dynamic and linear dynamic terms contribute the most to vertical variations of p', i.e., $\partial p'/\partial z$. The contribution to dw/dt in the vertical momentum equation from $\partial p'/\partial z$ has been found to be of approximately the same size as the contribution to dw/dt from buoyancy in supercell storms. Below cloud base, upward-directed dynamic pressure gradients due to the linear and nonlinear effects described above are responsible for forcing negatively buoyant air to the level of free convection. This forced lifting of stable air often gives rise to visually spectacular, laminar, striated low-level clouds in supercells, such as that appearing in Fig. 2.

4.2 Nonlinear effects

Let us take a closer look at the nonlinear dynamic term in (4.12), $\frac{1}{2}\zeta'^2$. It may seem strange that we are examining this term before looking at the linear dynamic term, but nonlinear effects frequently play an important role in the early evolution of a supercell updraft.

The $\frac{1}{2}\zeta'^2$ term leads to updraft splitting. Tilting of vortex lines leads to a couplet of vertical vorticity straddling the updraft (Fig. 8). Low pressure is found where $|\zeta|$ is large, regardless of the sign of ζ (i.e., elevated low pressure is produced in both the cyclonic *and* anticyclonic members of the vortex pair). Since $|\zeta|$ is a maximum aloft, p' is a minimum aloft on the flanks, and an upward-directed dynamic vertical pressure gradient arises on the flanks below the level of maximum $|\zeta|$ (e.g., 4–6 km). These dynamic pressure gradients on the opposing flanks cause the updraft to split. The splitting process may be observed within approximately 30–60 minutes of storm initiation, and may occur successively. Splitting also may be encouraged by rainwater and associated downdraft formation along the axis of the initial updraft, but a rainy downdraft is not needed to produce splitting.

Splitting leads to the formation of a pair of right- and left-moving storms. Vertical vorticity constantly is produced by tilting on the updraft flanks (where $\nabla_h w$ is largest), which leads to continuous propagation laterally (at right angles to the shear vector) as the right-moving (left-moving) updraft is continuously forced on its right (left) flank. The updraft of the right (left) mover acquires cyclonic (anticyclonic) rotation, and the downdraft of the right (left) mover acquires anticyclonic (cyclonic) rotation. The $\frac{1}{2}\zeta'^2$ term gives rise to lateral, off-hodograph propagation as long as ζ and w are not co-located.

4.3 Linear effects

Now let us take a closer look at the linear dynamic term, $2\frac{\partial \bar{\mathbf{v}}_h}{\partial z} \cdot \nabla_h w'$, in (4.12). High pressure is found upshear (not *upwind*) of the updraft and low pressure is found downshear (not *downwind*) of the updraft. The couplets of high and low pressure are aligned with the shear vector at that level (recall that the vorticity couplets were aligned normal to the shear). See Fig. 9.

When the vertical wind shear is unidirectional, the highs and lows are vertically stacked so that significant upward- or downward-directed vertical pressure gradients do not arise. Mirror-image, symmetrical right- and left-moving storms are produced, and these propagate off of the hodograph by way of the nonlinear $\frac{1}{2}\zeta'^2$ effects. When the shear vector veers with height (i.e., the hodograph turns clockwise), the high-low pressure couplets become stacked such that an upward-directed dynamic vertical pressure gradient exists on the right flank of the updraft and a downward-directed dynamic vertical pressure gradient exists on the left flank of the updraft. This leads to the suppression of the left-moving storm and enhancement of the right-moving storm. When the hodograph turns counterclockwise with height, the opposite is true—the left-mover is favored and the right-mover is suppressed.

Recall that for strongly curved hodographs, there is a large correlation between w and ζ. In the limit of a circular hodograph, the w and ζ fields are co-located (for practical purposes, but not exactly, because of buoyancy in the updraft); i.e., ζ is not centered on the flanks and the nonlinear term ($\frac{1}{2}\zeta'^2$) cannot lead to deviant propagation because the upward-directed vertical pressure gradient force due to the nonlinear term is strongest in the updraft *center*. But in this circular hodograph limit, linear contributions to $\partial p'/\partial z$ are strongest and the dynamic vertical pressure gradient is directed upward on the right flank; i.e., the a right-mover is enhanced on the right flank by the turning of the shear vector with height, and rightward propagation occurs. So for a straight hodograph, deviant propagation owes to the nonlinear term in (4.12), and for a curved hodograph, deviant propagation owes to the linear term in (4.12).

5 Further reading

Davies-Jones, R., 1984: Streamwise vorticity: The origin of updraft rotation in supercell storms. *J. Atmos. Sci.*, **41**, 2991–3006.

Davies-Jones, R., R. J. Trapp, and H. B. Bluestein, 2001: Tornadoes and tornadic storms. *Severe Local Storms, Meteor. Monogr.*, No. 50, 167–222.

Klemp, J. B., 1987: Dynamics of tornadic thunderstorms. *Ann. Rev. Fluid Mech.*, **19**, 369–402.

Klemp, J. B., and R. B. Wilhelmson, 1978: Simulations of right- and left-moving storms produced through storm splitting. *J. Atmos. Sci.*, **35**, 1097–1110.

Klemp, J. B., R. B. Wilhelmson, and P. S. Ray, 1981: Observed and numerically simulated structure of a mature supercell thunderstorm. *J. Atmos. Sci.*, **38**, 1558–1580.

Klemp, J. B., and R. Rotunno, 1983: A study of the tornadic region within a supercell thunderstorm. *J. Atmos. Sci.*, **40**, 359–377.

Markowski, P. M., 2002: Hook echoes and rear-flank downdrafts: A review. *Mon. Wea. Rev.*, **130**, 852–876.

Markowski, P. M., J. M. Straka, and E. N. Rasmussen, 2002: Direct surface thermodynamic observations within the rear-flank downdrafts of nontornadic and tornadic supercells. *Mon. Wea. Rev.*, **130**, 1692–1721.

Rotunno, R., 1981: On the evolution of thunderstorm rotation. *Mon. Wea. Rev.*, **109**, 577–586.

Rotunno, R., and J. B. Klemp, 1982: The influence of the shear-induced pressure gradient on thunderstorm motion. *Mon. Wea. Rev.*, **110**, 136-151.

Rotunno, R., and J. B. Klemp, 1985: On the rotation and propagation of simulated supercell thunderstorms. *J. Atmos. Sci.*, **42**, 271–292.

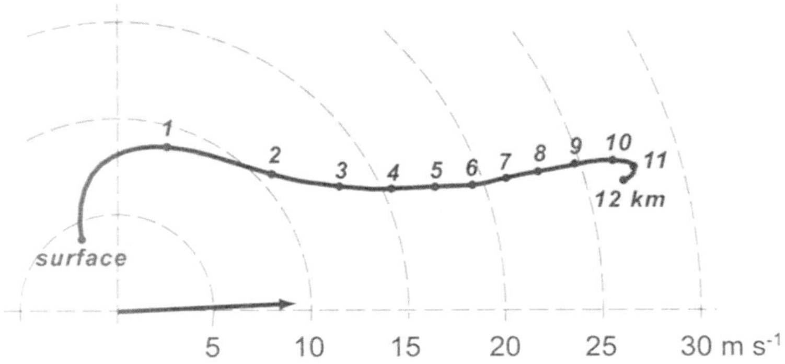

Figure 1. Composite hodograph obtained from the inflow environments of >400 supercell thunderstorms in the United States. Markings along the hodograph are placed at 1 km intervals, starting at the ground. The filled circle indicates the mean storm motion.

Figure 2. A midlevel mesocyclone is one of the defining characteristics of a supercell storm. Little imagination is needed to sense the cyclonic vertical vorticity associated with the storm updraft. Photograph by Paul Markowski on 13 June 1998 near Oklahoma City, Oklahoma.

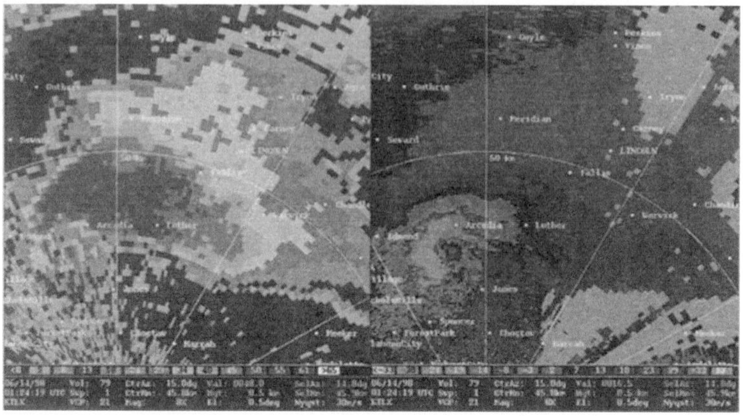

Figure 3. A hook echo and Doppler radial velocity couplet are the defining radar characteristics of supercells. The images above are reflectivity and radial velocity from the Oklahoma City radar at 0124 UTC 13 June 1998. This is the same storm appearing in Fig. 2, about 5 min after the photo was taken.

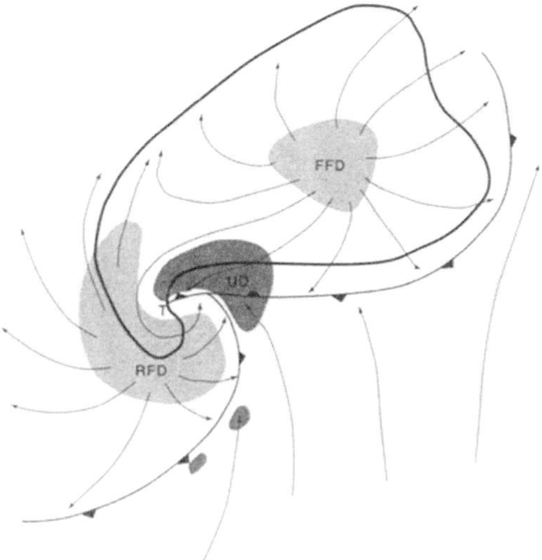

Figure 4. The Lemon and Doswell (1979) supercell conceptual model remains largely unmodified today. The closed, bold curve represents the generalized echo shape as viewed by radar. "UD" refers to the updraft location, "FFD" is the forward-flank downdraft, and "RFD" is the rear-flank downdraft. If a tornado is present, it typically is located near the "T." [Adapted from Lemon and Doswell (1979).]

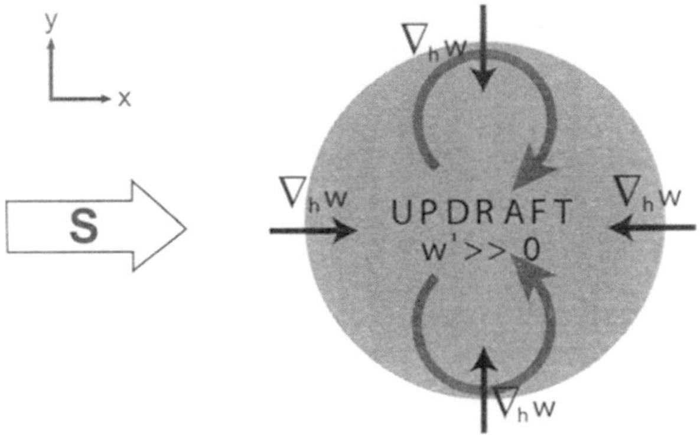

Figure 5. An $x - y$ cross-section of an updraft (gray) at midlevels, indicating the relationship between ζ' (its sense is given by the light blue arrows), vertical wind shear (**S**), and $\nabla_h w$ as a result of the $\mathbf{S} \times \nabla_h w' \cdot \mathbf{k}$ term in (3.4).

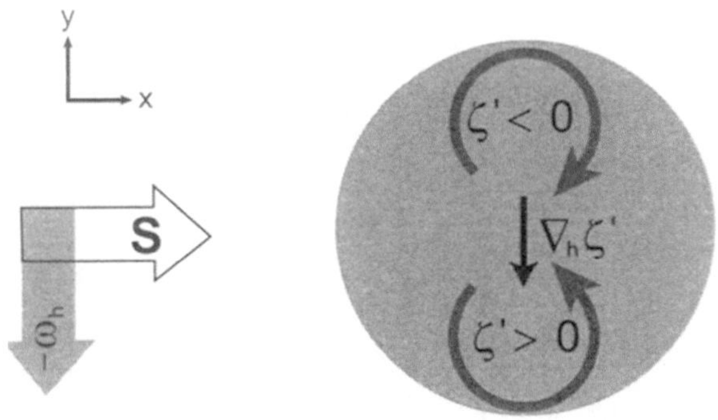

Figure 6. An $x - y$ cross-section of an updraft (gray) at midlevels, indicating the relationship between ζ' (its sense is given by the light blue arrows), **S**, $-\boldsymbol{\omega}_h$, and $\nabla_h \zeta'$ appearing in the $-(\bar{\mathbf{v}} - \mathbf{c}) \cdot \nabla_h \zeta'$ term in (3.4). The lateral shifting of the ζ' field within the updraft cross-section depends on the orientation of the storm-relative wind with respect to $-\boldsymbol{\omega}_h$ (and $\nabla_h \zeta'$, which points in the same direction as $-\boldsymbol{\omega}_h$). This orientation is related to whether the inflow horizontal vorticity is streamwise or crosswise.

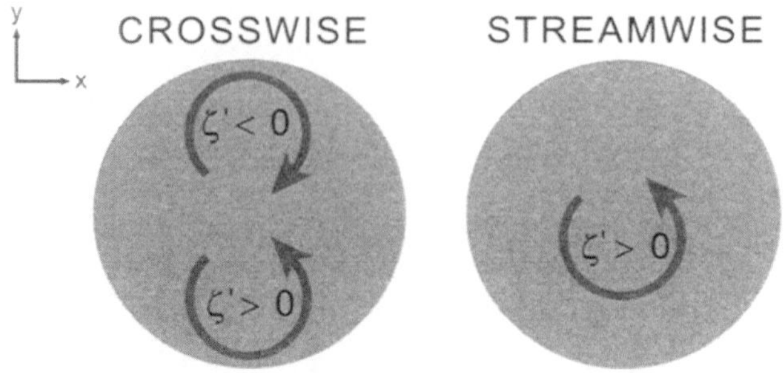

Figure 7. An $x-y$ cross-section of an updraft (gray) at midlevels, indicating the relationship between ζ' (its sense is given by the light blue arrows) and updraft maximum (assumed to be at the center of the updraft) depending on whether predominantly crosswise vorticity has been tilted or predominantly streamwise vorticity has been tilted.

NONLINEAR CONTRIBUTION TO DYNAMIC PRESSURE PERTURBATIONS

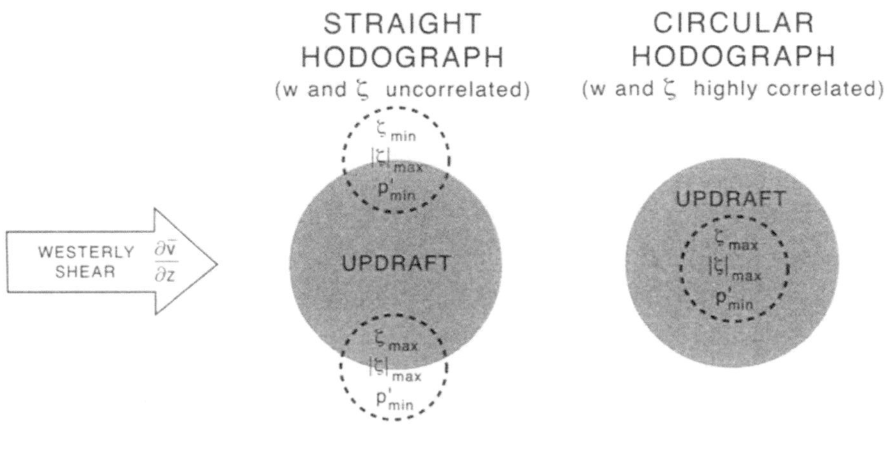

Figure 8. An $x-y$ cross-section of the ζ (dashed) and updraft (gray) fields at midlevels for a straight and curved hodograph. For a straight hodograph, the $\frac{1}{2}\zeta'^2$ term leads to an upward-directed dynamic vertical pressure gradient force on the storm flanks below the level of maximum ζ and minimum p' (associated with the maximum ζ). In the limit of a circular hodograph, the ζ and w fields are co-located, and the $\frac{1}{2}\zeta'^2$ term cannot lead to an off-axis, upward-directed, dynamic vertical pressure gradient force.

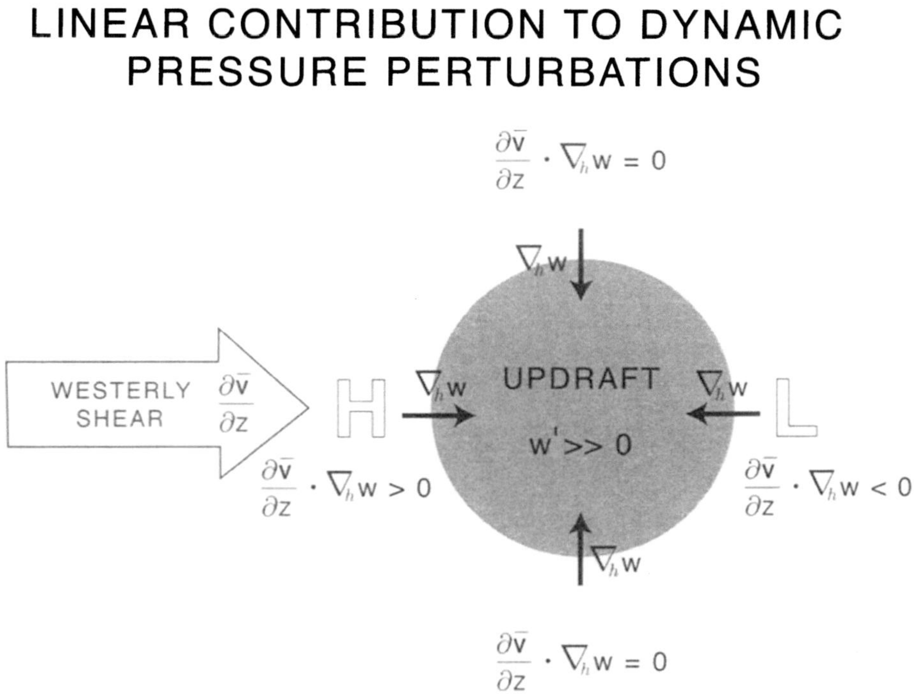

Figure 9. Pressure perturbations arising from the linear dynamic term, $2\frac{\partial \bar{\mathbf{v}}}{\partial z} \cdot \nabla_h w'$. A couplet of high and low p' is aligned with the shear vector and straddles the updraft at any particular altitude. The amplitude of p' is proportional to the magnitudes of the horizontal w gradient (strong updrafts tend to also have large $\nabla_h w'$) and the vertical wind shear.

Tornadoes and Tornadogenesis

Paul Markowski

Department of Meteorology, Penn State University, University Park, Pennsylvania, USA

Abstract. Tornado structure and dynamics are reviewed. The conditions leading to tornadogenesis also are discussed, as are some of the latest forecasting techniques used to discriminate between tornadic and nontornadic supercells.

1 General characteristics of tornadoes

Tornadoes are violently rotating columns of air, usually associated with a swirling cloud of debris or dust near the ground and a funnel-shaped cloud extending downward from the base of the parent cumulonimbus updraft. Most tornadoes have winds <50 m s^{-1}, although the strongest 10%, with winds >80 m s^{-1}, account for a disproportionate number of fatalities (~90%). Vertical vorticity, ζ, has magnitudes on the order of 1 s^{-1}, given typical tornado diameters on the order of 100 m. Vertical velocities also may be intense, especially just above the surface layer, with values >75 m s^{-1} possible. Radial inflow velocities typically range from 20–60 m s^{-1}. Most tornadoes persist <10 min. Tornado size (diameter) and longevity generally increase with intensity, although many exceptions have been observed.

Roughly 99% of all tornadoes in the northern hemisphere rotate counterclockwise; for this reason, many early investigators believed that the Coriolis force played a direct role in tornadogenesis. But the role of the Coriolis force is only indirect. Cyclonic tornadoes are favored because right-moving supercells tend to be favored over left-movers in typical supercell environments, owing to clockwise-turning hodographs in most supercell environments. Hodographs tend to turn clockwise with height at low levels because of the influence of surface drag and the Coriolis force (the Ekman spiral is a curved hodograph that can be derived analytically when surface drag and rotation effects are considered). Therefore, only *indirectly* by way of supercell updraft dynamics, the Coriolis force plays a role in the cyclonic tornado bias.

Most significant tornadoes (which I arbitrary define as those with winds ≥50 m s^{-1} *or* those that persist >5 min) are associated with supercell storms, although weak tornadoes occasionally occur in nonsupercell storms. Virtually all strong and violent tornadoes (those ≥F2) are associated with supercells.

2 Structure and dynamics

2.1 Flow regions

Our present understanding of the morphology and dynamics of tornadoes has been gained by photogrammetric studies, laboratory experiments, and numerical simulations. The flow in tornadoes can be partitioned into five regions (Fig. 1):

Outer region (I): Flow moves in response to the large vorticity in the core (Region II) and to the positive buoyancy and vertical pressure gradient associated with the updraft aloft (Region V). Flow in this region generally conserves its angular momentum, therefore air spins faster as it approaches the tornado axis.

Core (II): Extends outward from the tornado axis to the radius of maximum winds. It often contains a funnel cloud and a column of dust and debris from the ground. Because of the intense rotation in this region, air parcels approximately are in cyclostrophic balance; i.e., the apparent centrifugal force owing to the large curvature and high speed of the flow is balanced by a horizontal pressure gradient force directed inward toward the axis.

Corner (III): The part of the boundary layer in which the flow turns upward from a primarily horizontal direction to a primarily vertical direction.

Boundary layer flow (IV): This layer is anywhere from 10 to 100 m deep and is turbulent owing to the interaction of the flow with the ground. Friction precludes cyclostrophic balance, and radial inflow toward the axis is produced. The radial inflow and convergence of angular momentum (which is nearly conserved) give rise to intense wind speeds. Thus, paradoxically, the presence of friction can actually intensify the vortex.

Rotating updraft (V): The parent, storm-scale, rotating updraft situated above the tornado.

2.2 Wind and pressure profile

Within the core region, the tornado can be approximated to first-order as a Rankine vortex; i.e., constant angular velocity (v/r, where v is the tangential wind speed and r is the distance from the tornado center) is found within the radius of maximum tangential winds, and outside of the radius of maximum tangential winds, angular momentum (vr) is constant:

$$v = \begin{cases} v_{max} r / r_{max}, & r \leq r_{max} \\ v_{max} r_{max} / r, & r > r_{max} \end{cases} \tag{2.1}$$

where r_{max} is the radius of the maximum tangential wind speed, v_{max}.

The flow within the tornado core is approximately in cyclostrophic balance. In this case, the momentum equation in natural coordinates (or the v equation in cylindrical coordinates) becomes

$$\frac{v^2}{r} = \frac{1}{\rho}\frac{\partial p}{\partial r} \tag{2.2}$$

$$\frac{\rho v^2}{r} = \frac{\partial p}{\partial r}. \tag{2.3}$$

To obtain the pressure, we integrate (2.3) (and assume p is a function of r only, i.e., $\partial p/\partial r = dp/dr$):

$$\int_{p(r)}^{p_\infty} dp = \int_r^\infty \frac{\rho v^2}{r} dr, \tag{2.4}$$

where $p(r)$ is the pressure at radius r from the vortex center and $p = p_\infty$ is the ambient pressure value found at an infinite distance from the vortex. We will assume ρ = constant. For $r > r_{max}$ [using (2.1)], we have

$$p_\infty - p(r) = \int_r^\infty \rho \left(\frac{v_{max} r_{max}}{r}\right)^2 \frac{dr}{r} \tag{2.5}$$

$$= \rho v_{max}^2 r_{max}^2 \int_r^\infty \frac{dr}{r^3} \tag{2.6}$$

$$= \rho v_{max}^2 r_{max}^2 \left[-\frac{1}{2r^2}\right]_r^\infty \tag{2.7}$$

$$= \frac{1}{2}\rho v_{max}^2 \frac{r_{max}^2}{r^2}; \tag{2.8}$$

therefore,

$$p(r) = p_\infty - \frac{1}{2}\rho v_{max}^2 \frac{r_{max}^2}{r^2} \quad \text{for } r > r_{max}. \tag{2.9}$$

For $r \le r_{max}$ [using (2.1)], we have

$$p_\infty - p(r) = \int_r^{r_{max}} \rho \left(\frac{v_{max} r}{r_{max}}\right)^2 \frac{dr}{r} + \int_{r_{max}}^\infty \rho \left(\frac{v_{max} r_{max}}{r}\right)^2 \frac{dr}{r} \tag{2.10}$$

$$= \frac{\rho v_{max}^2}{r_{max}^2} \int_r^{r_{max}} r\, dr + \rho v_{max}^2 r_{max}^2 \int_{r_{max}}^\infty \frac{dr}{r^3} \tag{2.11}$$

$$= \frac{\rho v_{max}^2}{r_{max}^2} \left[\frac{r^2}{2}\right]_r^{r_{max}} + \rho v_{max}^2 r_{max}^2 \left[-\frac{1}{2r^2}\right]_{r_{max}}^\infty \tag{2.12}$$

$$= \frac{1}{2}\rho v_{max}^2 - \frac{1}{2}\rho v_{max}^2 \frac{r^2}{r_{max}^2} + \frac{1}{2}\rho v_{max}^2 \tag{2.13}$$

$$= \rho v_{max}^2 \left(1 - \frac{1}{2}\frac{r^2}{r_{max}^2}\right); \tag{2.14}$$

therefore,

$$p(r) = p_\infty - \rho v_{max}^2 \left(1 - \frac{1}{2}\frac{r^2}{r_{max}^2}\right) \quad \text{for } r \le r_{max}. \tag{2.15}$$

The maximum pressure deficit, $p' = p_\infty - p(r)$, is obtained at the vortex center, where $r = 0$:

$$p'_{max} = \rho v_{max}^2. \tag{2.16}$$

So for $\rho \sim 1$ kg m^{-3} and a weak tornado ($v_{max} \sim 25$ m s^{-1}), $p' \sim 6$ mb. This pressure deficit is only enough to lower the cloud base by approximately 60 m. Hence, a debris cloud may be present at the surface, but only a short funnel cloud would be visible aloft. If the tornado intensifies to $v_{max} = 70$ m s^{-1}, $p' \sim 50$ mb, and the cloud base lowers by 500 m, which likely would be near to the ground. The funnel cloud is not a material surface—it descends as air is continually flowing upward through it.

2.3 Swirl ratio

The dynamics of a tornado vortex depend strongly on a dimensionless parameter called the swirl ratio, S,

$$S = \frac{r_o \Gamma}{2Q} = \frac{v_o}{w_o}, \tag{2.17}$$

where Γ is the circulation about the central axis at radius $r = r_o$, Q is the rate of volume flow through the top of the tornado chamber or simulation domain (the importance of the swirl ratio has been identified from laboratory and numerical experiments), v_o is the tangential velocity at r_o, and w_o is the mean vertical velocity at the top of the domain (near the top of the tornado).

For small values of S (e.g., <1), the flow is dominated by updraft rather than by rotation. At any given level above the boundary layer, the pressure rises as the flow approaches the axis because the air parcels must decelerate. Thus, the boundary layer "separates" and any rotation that exists in the entering stream is carried aloft before reaching the axis. As S is increased, the rotation in the outer flow makes the pressure gradient in the direction of the flow favorable so that surface flow can penetrate to the axis—a "one-celled" vortex is formed. "One-celled" refers to the fact that the vortex contains updraft throughout. As S is increased further, a central downdraft forms and a "two-celled" vortex results. The large swirl ratio (e.g., >2) is associated with a low-level pressure deficit that is so large that it causes a net axial downflow. "Two-celled" refers to the fact that a central downdraft is surrounded by updraft in this case.

The transition from a one-celled vortex to a two-celled vortex is termed "vortex breakdown." When S is very large, the central downdraft reaches the ground. The interface between the rapidly rotating inflow and weakly rotating downdraft air at the surface is unstable, and multiple vortices are formed. The multiple vortices rotate about the parent tornado circulation.

Many tornadoes have been observed to transition from roughly one-celled to two-celled tornadoes and exhibit vortex breakdown. In fact, most strong tornadoes are believed to exhibit multiple vortices. Fujita called these subvortices "suction spots," because they were associated with localized intense damage within the broader debris swath of the tornado (presumably owing to the superposition of the velocity field associated with the small-scale subvortex and the larger-scale tornado).

3 Tornadogenesis

3.1 Vertical vorticity generation at the ground

A tornado, by definition, is a violently rotating column of air in contact with the ground. In other words, large vertical vorticity must be generated at the ground in order for a tornado to develop. If preexisting vertical vorticity is lacking near the ground, then vorticity stretching near the ground is initially negligible and vertical vorticity first must arise either from the tilting of horizontal vorticity, or from advection toward the surface from aloft. Tilting by the horizontal vertical velocity gradients associated with an updraft alone is not effective at producing vertical vorticity near the surface because air is rising away from the surface as horizontal vorticity is tilted into the vertical. But if a downdraft is involved in the tilting process, then vertical vorticity can arrive at the surface as it is produced via tilting, where it subsequently can be stretched to form a tornado. For these reasons, it has been argued that a downdraft is needed for tornadogenesis when preexisting rotation is absent near the ground.

There is a substantial body of observational and numerical modeling evidence that supports the theoretical notion that a downdraft is important for tornadogenesis (again, when preexisting rotation is absent at the surface). For example, there are nearly countless observations of rear-flank downdrafts (RFDs) and associated hook echoes and "clear slots" in close proximity to tornadoes. Furthermore, trajectory analyses in a limited number of observed supercells indicate that at least some of the air entering the tornado passes through the RFD prior to entering the tornado. This finding also is implied by observations that, at the time of tornadogenesis, the RFD typically occludes the mesocyclone at low-levels. Numerical simulation results also have emphasized the importance of the RFD and have shown similar trajectories of air parcels entering modeled vortices resembling tornadoes.

When there is preexisting rotation at the surface, a downdraft such as the RFD clearly is not needed for tornadogenesis. In these cases, near-ground stretching (convergence) alone can amplify ζ to tornado intensity. It seems as though waterspouts, landspouts, and perhaps most other geophysical vortices commonly arise in this manner.

3.2 The forecasting dilemma

Environmental variability Although supercells are relatively easy to anticipate, predicting which supercells will spawn tornadoes is one of the most arduous tasks facing operational meteorologists and researchers alike. In the first Doppler radar studies conducted in the 1970s in Oklahoma, it was found that roughly 50–60% of all supercells spawned tornadoes. But once a Doppler radar network became installed in the United States (mid-1990s), mesocyclones and supercells were detected nationwide at an increasing rate, and the percentage of supercells associated with tornadoes declined significantly. Today it is widely accepted that as few as 10–20% of supercells produce tornadoes. Thus, if a tornado warning is issued for each supercell detected by Doppler radar, one might expect an 80–90% false alarm rate.

In Fig. 2, a radar display during a severe weather outbreak in central Oklahoma on 4 October 1998 is shown. Numerous supercells are evident, but what is disturbing is the differences in appearance of the supercells over small horizontal distances. Why do the storms look so different? Why are some tornadic and others not? One might reasonably ask whether there are any differences between the mesoscale environments of tornadic and nontornadic supercells.

It has been found that storm environments can contain so much small-scale variability that it is often difficult to even define what one means by "storm environment." In Fig. 3, 0–1 km storm-relative helicity (SRH) values are shown in a small 23×30 km^2 region during a field experiment, during which our ability to sample small-scale details far surpassed what we can routinely observe. Generally speaking, some of the environmental variability like that revealed in Fig. 3 (e.g., in terms of parameters like SRH, etc.) seems to be associated with mesoscale phenomena that can marginally be resolved by the conventional observing systems (e.g., outflow boundaries, topographic features). But it also is clear that much of the environmental variability is associated with boundary layer convection (e.g., cellular convection, rolls), which is ubiquitous to boundary layers during sunny, daytime conditions. What is unsettling is that we are presently ignorant of the effects such small-scale variability have on convective storms, and perhaps tornadogenesis. Perhaps some of the finescale variability may explain why some supercells produce tornadoes and others do not in *seemingly* similar environments, such as in Fig. 2—the environments only seem similar on most days because conventional observing networks (with less resolution than the observations in Fig. 3) cannot sample most of the variability present on the mesoscale. Thus, tornadoes often appear to be "mesoscale accidents." Of course, on some days, tornado production in supercells is nearly widespread (e.g., "outbreak days," like 3 May 1999 in central Oklahoma), which may be an indication that the larger-scale conditions may be so conducive to tornado formation in such cases that mesoscale variability becomes unimportant in the tornado versus no-tornado question. Such "outbreak days" are relatively rare, however.

Although much of the small-scale environmental variability may not be directly observable, locations where horizontal vorticity and SRH may be locally enhanced may be inferred from the detection of preexisting boundaries. Such boundaries may include outflow boundaries, warm fronts, inland sea breezes, etc., and may be visible in satellite imagery as lines of cumulus clouds or in radar data as fine-lines. The thermal contrasts associated with such features generate horizontal vorticity baroclinically (Fig. 4). Horizontal vorticity is generated baroclinically with an orientation parallel to the isotherms (density isopleths), with cold air on the right if looking down the vorticity vector. This "baroclinic" horizontal vorticity may augment the horizontal vorticity present on the larger scale that is due to the mean vertical wind shear, and SRH also may be enhanced by such a thermal boundary if some of the vorticity generated has a streamwise component. Supercells that encounter regions where horizontal vorticity and SRH are locally augmented have been found to have a greater likelihood to produce tornadoes than supercells occurring in "pristine" warm sectors.

Radar discrimination between tornadic and nontornadic supercells It is very difficult to distinguish between tornadic and nontornadic supercells in Doppler radar data. Except in rare circumstances, radars only detect tornado parent circulations (i.e., mesocyclones)—they cannot resolve tornadoes themselves.

A disturbing collage of supercells as viewed in radar reflectivity data is presented in Fig. 5. No obvious differences are apparent between the tornadic and nontornadic supercells. In fact, some of the more impressive looking hook echoes are not associated with tornadoes. In radial velocity data, the strongest midlevel mesocyclones are not necessarily the ones associated with tornadoes. Tornadoes occur over a broad range of midlevel mesocyclone intensities.

The latest in discriminating between tornadic and nontornadic supercells Finally, it may be beneficial to discuss some of the latest attempts to improve forecasters' abilities to discriminate between tornadic and nontornadic supercells. The two parameters that seem to show the most promise, assuming that conditions favor "surface-based" supercells,[1] are (1) low lifting condensation level (LCL) heights and (2) large low-level (0–1 km) wind shear and SRH.

When LCLs are low, supercells are more likely to produce significant tornadoes than when LCL heights are high. This may be due to the fact that low LCLs inhibit the production of cold outflow, and excessively cold outflow seems to be detrimental to the production of tornadoes by supercell thunderstorms. It has recently been found that RFDs that arrive at the ground with only small (1–2 K) temperature deficits may be more favorable for tornadogenesis than RFDs that arrive at the ground very cold (>5 K temperature deficits). The precise dynamical importance of RFD thermodynamic traits has not yet been confirmed. One possibility is that the RFD air parcels entering the tornado or incipient tornado (previous studies have found that at least some of the air entering tornadoes comes from the RFD), if too cold and stable, inhibit lifting and convergence, thereby limiting vorticity stretching.

4 Further reading

Davies-Jones, R., R. J. Trapp, and H. B. Bluestein, 2001: Tornadoes and tornadic storms. *Severe Local Storms, Meteor. Monogr.*, No. 50, 167–222.

Markowski, P. M., 2002: Hook echoes and rear-flank downdrafts: A review. *Mon. Wea. Rev.*, **130**, 852–876.

Markowski, P. M., E. N. Rasmussen, and J. M. Straka, 1998: The occurrence of tornadoes in supercells interacting with boundaries during VORTEX-95. *Wea. Forecasting*, **13**, 852–859.

Markowski, P. M., J. M. Straka, E. N. Rasmussen, and D. O. Blanchard, 1998: Variability of storm-relative helicity during VORTEX. *Mon. Wea. Rev.*, **126**, 2959–2971.

[1]Supercells undercut by shallow, stable boundary layers ("elevated supercells") posed little tornado threat.

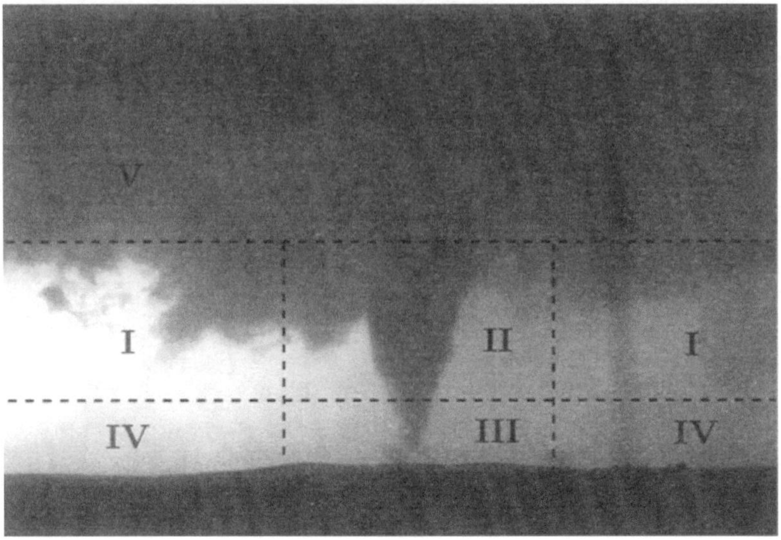

Figure 1. The five regions of a tornado vortex (see text for description). Photo by P. Markowski.

Markowski, P. M., J. M. Straka, and E. N. Rasmussen, 2002: Direct surface thermodynamic observations within the rear-flank downdrafts of nontornadic and tornadic supercells. *Mon. Wea. Rev.*, **130**, 1692–1721.

Markowski, P. M., J. M. Straka, and E. N. Rasmussen, 2003: Tornadogenesis resulting from the transport of circulation by a downdraft: Idealized numerical simulations. *J. Atmos. Sci.*, **60**, 795–823.

Markowski, P. M., C. Hannon, J. Frame, E. Lancaster, A. Pietrycha, R. Edwards, and R. Thompson, 2003: Characteristics of vertical wind profiles near supercells obtained from the Rapid Update Cycle. *Wea. Forecasting*, **18**, 1262–1272.

Thompson, R., R. Edwards, J. Hart, K. Elmore, and P. Markowski, 2003: Close proximity soundings within supercell environments obtained from the Rapid Update Cycle. *Wea. Forecasting*, **18**, 1243–1261.

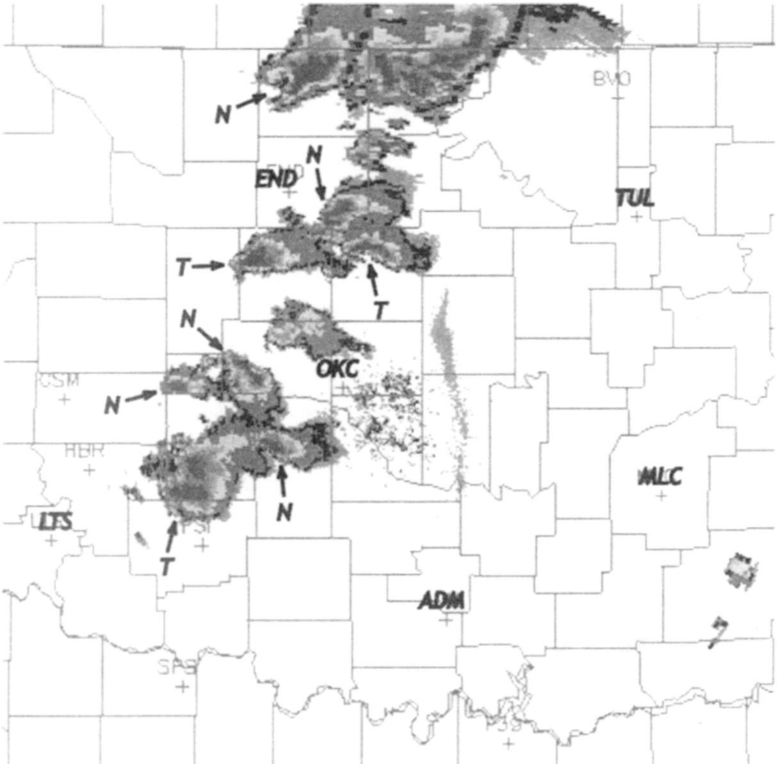

Figure 2. Radar image from 4 October 1998 during a rare fall tornado outbreak in Oklahoma. Over two dozen tornadoes were reported, yet most of the supercells were nontornadic. In this single image, numerous supercells are present and all of them have at least some differences in their reflectivity appearance. Tornadic supercells are indicated with a "T," and nontornadic supercells are indicated with an "N." Why the differences? What role have environmental inhomogeneities played?

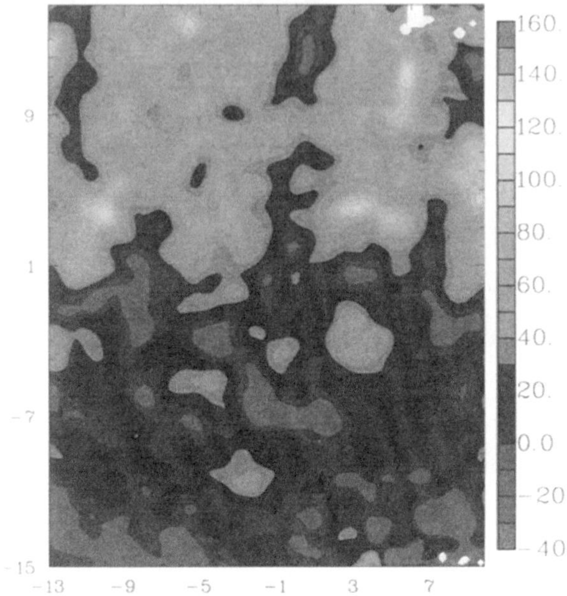

Figure 3. Example of environmental variability. Storm-relative helicity (SRH) values ($m^2\ s^{-2}$) have been analyzed in the 0–1 km layer in northwestern Oklahoma at 2015 UTC 12 June 2002, determined from a multiple-Doppler radar wind synthesis. Labels on the axes are horizontal distances in km. An outflow boundary extends from west to east across the middle of the domain, with values north of it being larger than values south of the boundary. Variability is present on smaller scales too, probably as a result of boundary layer convective rolls. How would we define the "environment" of a storm that would develop in this region?

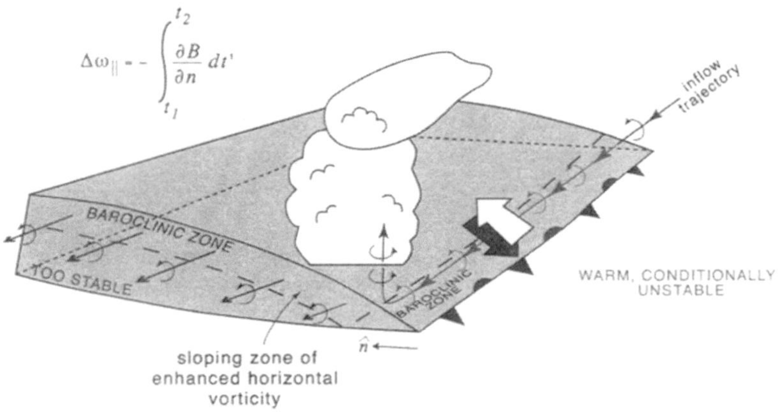

Figure 4. Schematic illustrating how baroclinity associated with a thermal boundary can augment the "ambient" horizontal vorticity associated with the mean vertical wind shear. A barocline generates horizontal vorticity parallel to the density isopleths, with cold air on the right, looking down the vorticity vector.

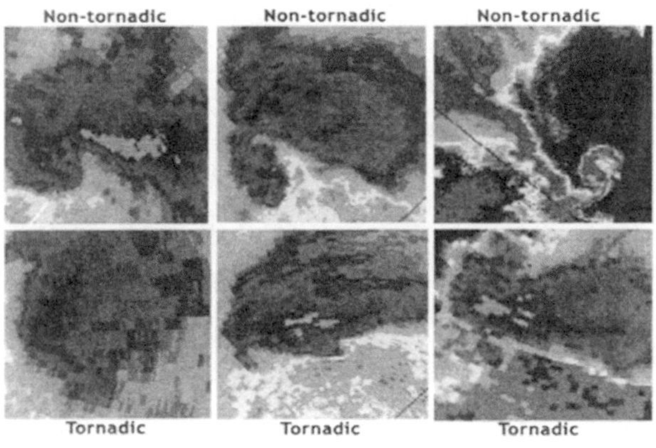

Figure 5. A sample of hook echoes associated with both tornadic and nontornadic supercells. The hook echoes associated with tornadic supercells are as they appeared 5 min or less prior to tornadogenesis. No obvious differences are apparent between the hook echoes associated with the tornadic and nontornadic supercells.

Dynamical Aspects of Topography: The Role of Obstacles

Reinhold Steinacker

Institute of Meteorology and Geophysics, University of Vienna, Austria

Abstract. In this document the dynamical aspects of topography is being discussed and the consequences for forcing or damping convection. Besides general flow modification due to topography foehn and Bora are discussed, followed by a chapter on fronts and orography.

1 Flow over versus flow around an obstacle

The simplest approach to study the role of mountains with regard to flow modification is to consider the exchange between kinetic and potential energy. For a single parcel in motion this can be done easily. If we know the initial speed of the parcel against the obstacle and the (relative) height of the obstacle we can solve the problem (see Figure 1.1).

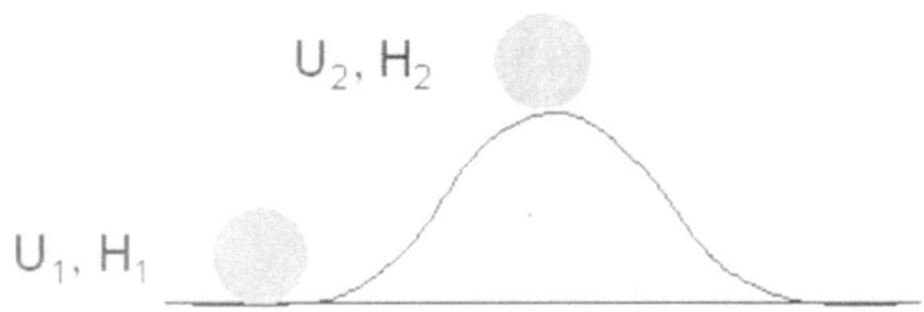

Figure 1. Schematic view of a parcel climbing a mountain.

The initial kinetic energy of the parcel is $\frac{MU_1^2}{2}$, the initial potential energy is MgH_1, with M being the mass of the parcel, U_i the initial speed towards the obstacle, g the acceleration of gravity and H_l the initial height of the parcel. To allow the parcel to climb the whole mountain, the initial kinetic energy must be at least $Mg(H_2 - H_1)$. If we look at the ratio between both

$$\frac{MU_1^2}{2Mg(H_2 - H_1)} = \frac{U_1^2}{2g(H_2 - H_1)} = \frac{U_1^2}{2gH} \succ 1, \quad (1.1)$$

where we denote the relative height of the mountain by H, this ratio must exceed unity, to allow the parcel climbing over the obstacle. If we want to apply this scheme to the atmosphere, we have to consider several differences and difficulties. First, an air parcel cannot be treated separately, it is always embedded in and interacts with the surrounding air. Second, even if we keep the thermal energy constant, i. e. we consider an isentropic process, a third form of energy comes into the game, the compression energy. To keep the ratio simple we may, however, define a "reduced" potential energy by using a "reduced" acceleration of gravity, which takes into account for the ambient static stability of the air:

$$g' = g\frac{1}{\Theta}\frac{\partial\Theta}{\partial z}\frac{H}{2} = N^2\frac{H}{2}, \qquad (1.2)$$

with Θ being the potential temperature and N the Brunt-Vaisala frequency of the ambient air. Substituting (1.2) into (1.1), and taking the square root of the ratio we receive

$$\frac{U_1}{NH} \equiv Fr \succ 1 \qquad (1.3)$$

as a criterion (Fr = Froude Number) for an air parcel to be able to climb the obstacle. In the atmosphere there is a limitation to the application. Due to a wave motion aloft, the related pressure perturbation may give an additional contribution to be considered (see, e. g. Smith, 1990). A further mayor difficulty, when applying the Froude number to the atmosphere is to find the adequate value for U_I , as the pressure perturbation due to the flow impinging the mountains may extend far upstream and influence the flow speed. Furthermore it is often difficult to decide which stability (dry or moist) is to be taken for the Froude number. Nevertheless, as a rough estimate, more qualitative than quantitative, the Froude number approach is certainly useful.

2 Foehn and Bura

When an air parcel is lifted on the windward side of an obstacle, cloud formation and eventually precipitation will be initiated, if the moisture content is sufficient. Furthermore, if the atmosphere is conditionally unstable, convection will be initiated. Hence, mountain ranges, oriented roughly perpendicular to the mean flow will in general have a pronounced humid windward and a dry leeward side. Examples for this are the Rocky Mountains, the southern Andes, the Scandinavian mountains or the New Zealand Alps. Even the rather small volcanic islands in the trade wind zone, like in Hawaii or the Canary islands do show such contrasts.

For the European Alps, which are mostly West – East oriented, there is no general wet (windward) or dry (leeward) side. There is a permanent change within the baroclinic wave regime of the Westerlies between a more southerly and a more northerly flow against the obstacle. The common flow across the Alps, the foehn flow, has aroused interest among meteorologists long ago. One of the basic questions was, why the air on the windward side is descending even if the air is warm - as compared to the windward side. The general cause of warming was first explained by Hann, 1866 through applying moist thermodynamics to the atmosphere, see Figures 2 and 3.

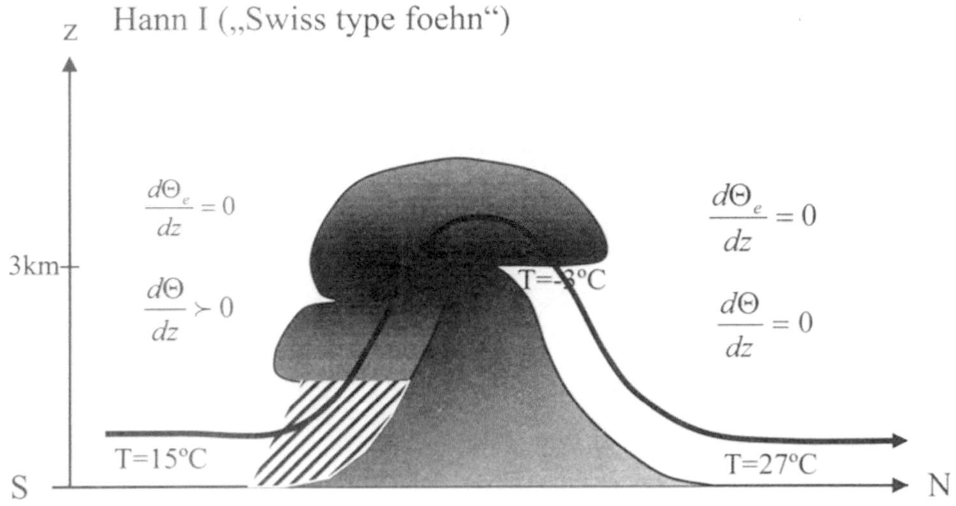

Figure 2. Schematic picture of a moist flow across a mountain ridge with windward precipitation and leeside foehn.

In Figure 2 the classical text book version of a foehn flow is presented. On the windward side the air is lifted until water vapour saturation is reached. The further ascent from the condensation level is moist adiabatic up to the crest of the mountain. The descent on the lee side is dry adiabatic the whole depth, leading to a considerable temperature surplus on the leeward side, as compared to the windward side. Hann was aware of the fact, that on many foehn occasions there is no precipitation on the windward side of the Alps, so that latent heat input cannot be the cause for the still present leeside warming. He presented the second type of foehn, which brings warmer air to the lee side simply by dry adiabatic descent, whereas at the windward side stably stratified air stays blocked. The second case (see Figure 3) was forgotten or being neglected for a long time since its discovery by Hann. Due to climatology the first type of foehn is typical for the Western (Swiss) Alps, whereas the latter is more common over the Eastern (Austrian) Alps.

The question was still open, why the warm air is becoming accelerated towards the leeside foot of the mountains, in contrary to the Archimedes' principle. The answer for this question took fairly long, until the dynamics of mountain waves was understood. The

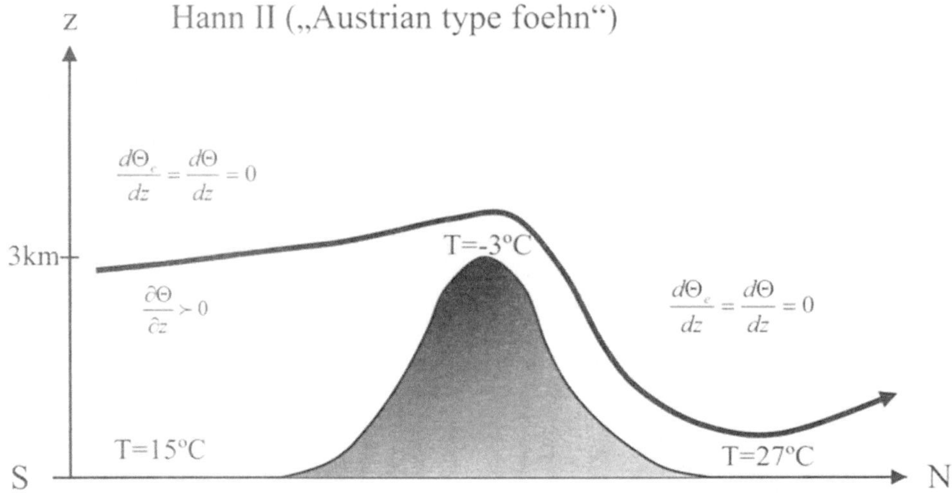

Figure 3. Schematic picture of a dry foehn flow across a mountain ridge without windward precipitation. The cross mountain temperature difference is solely due to a stably stratified blocked air on the windward side.

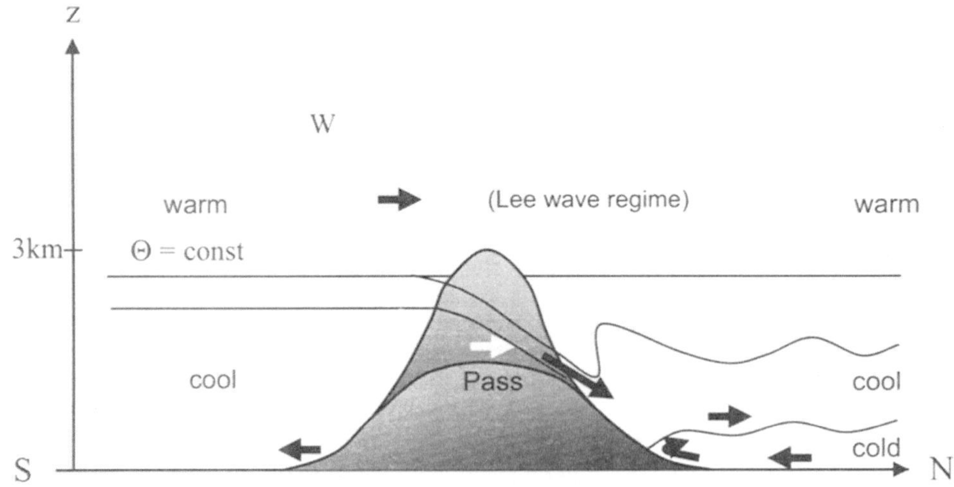

Figure 4. Schematic picture of a shallow foehn/ gap flow through a mountain pass.

descending warm air is forced downward due to the pressure perturbation of the wave regime. An additional, very important contribution to the understanding of the downward acceleration on the lee side during foehn was discovered during ALPEX, 1982, when the importance of passes in the trans - mountain flow (shallow foehn, gap flow, see Figure 4) and the essential role of shallow cold air pools on the lee side was recognized. Passes in a mountain ridge allow a small part of the windward blocked air to flow at comparatively low levels towards the lee side. This shallow foehn or gap flow is gravitationally driven and can be successfully described by hydraulic theory (see e. g. Mayr, 2003). The shallow foehn has the same characteristics as Bora, the cold gusty downslope windstorm on the Eastern Adriatic coast, see e. g. Barry, 1992. But why produces the (shallow) foehn a warming during its onset, in contrast to the cooling during the onset of Bora? The explanation to this is due to the existence of a shallow cold air pool on the lee side of mountains, which may be suddenly removed mechanically and being replaced by the cool – albeit warmer - air of the shallow foehn (Steinacker, 2006). On some occasions during summer time, when no cold air pool is present, during the shallow foehn onset e. g. at Innsbruck, a cooling is observed (see e. g. Fliri, 1975).

The foehn over complex topography with passes in the crest is hence a three layer process, the cold air pool with its own dynamics, the shallow bora – type foehn below crest height and the deep foehn with lee waves aloft. From a meteorological point of view the shallow foehn should be called a "masked Bora".

The life cycle of an Alpine foehn event is typically connected with a passage of a baroclinic wave from West to East. Significant stages of such a foehn cycle are presented in the following Figures 5 to 16. During the first stage (Figures 5 and 6) a high pressure and a ridge aloft is sitting over the Alps. There is no airmass contrast between both sides of the mountains. Besides thermally driven circulations no synoptic flow is present. During stage 2 (Figures 7 and 8) the center of the surface high is moving eastward. The pressure to the West and North of the Alps is falling so that an ageostrophic motion towards Northwest is initiated. Mass continuity requires a subsiding motion North of the Alps so that a temperature difference is built up across the ridge. At crest height there is either little wind or a flow from West to Northwest.

During the 3rd phase of the foehn cycle (Figures 9 and 10) the pressure difference accros the Alps is further increasing so that supported by the increasing air mass contrast between both sides of the mountains the shallow foehn starts. The wind at crest height is typically still Westerly, i. e. parallel to the crest. Due to the approaching surface low and the turning of the upper level flow towards a Southwesterly direction, at phase 4 (Figures 11 and 12) the deep foehn starts over the Alps. Upper level lee waves are produced. At low levels still the shallow foehn with its Bora like feature is present.

When the cold front of the baroclinic system approaches the Alps at phase 5 (Figures 13 and 14) of the foehn cycle, the foehn flow cannot reach the surface of the Northern foot

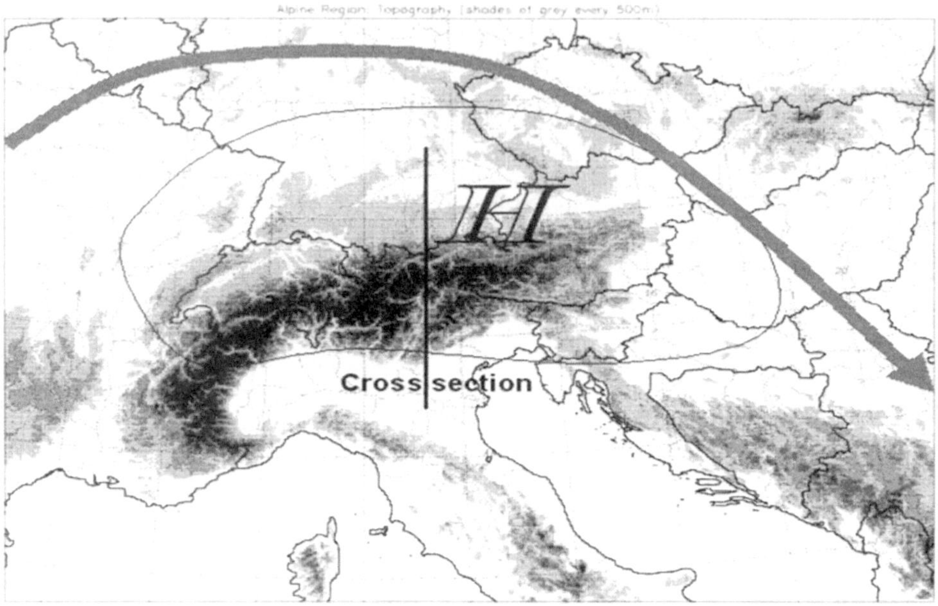

Figure 5. Phase 1 of a Foehn cycle, schematic surface isobars (thin) and upper tropospheric flow (shaded bold), further explanation see text

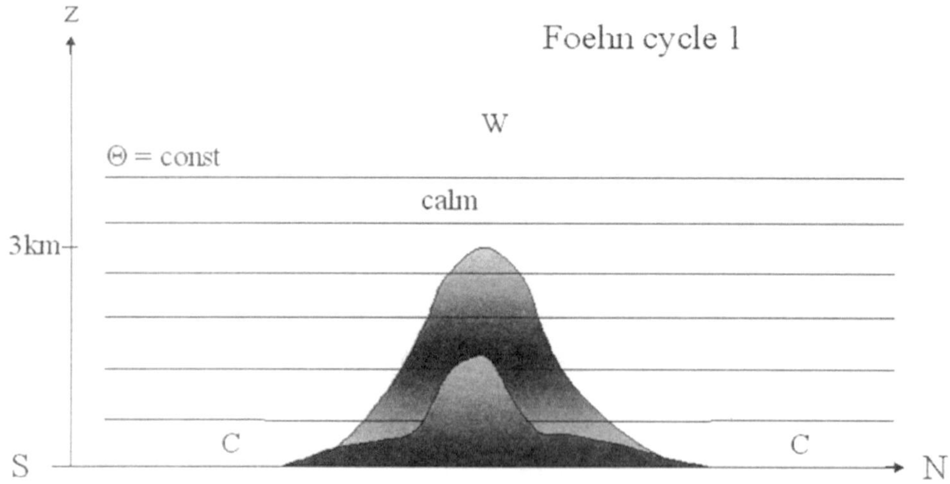

Figure 6. Isentropes along the cross section of figure 5

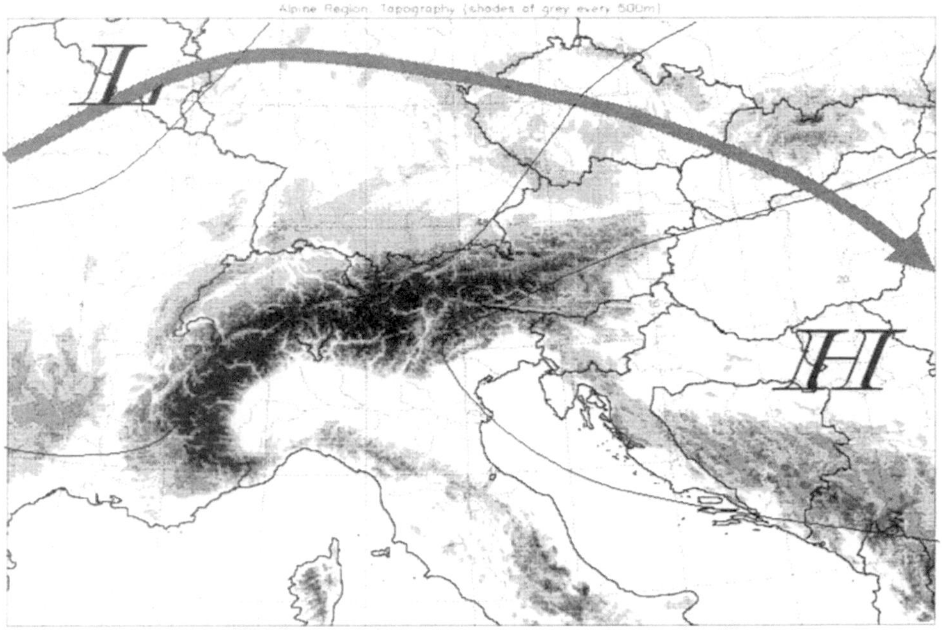

Figure 7. same as Figure 5 but for phase 2

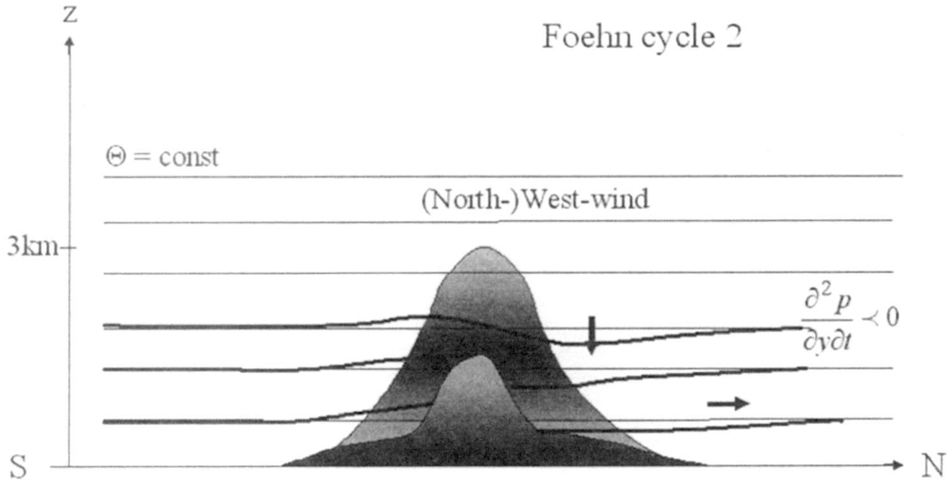

Figure 8. same as Figure 6 but for phase 2. Bold lines are perturbed isentropes

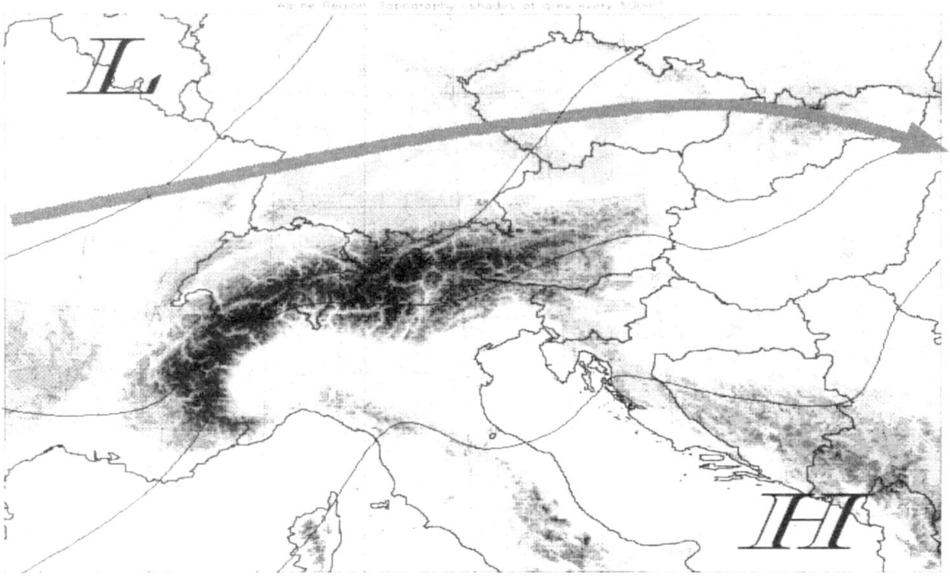

Figure 9. same as Figure 5 but for phase 3

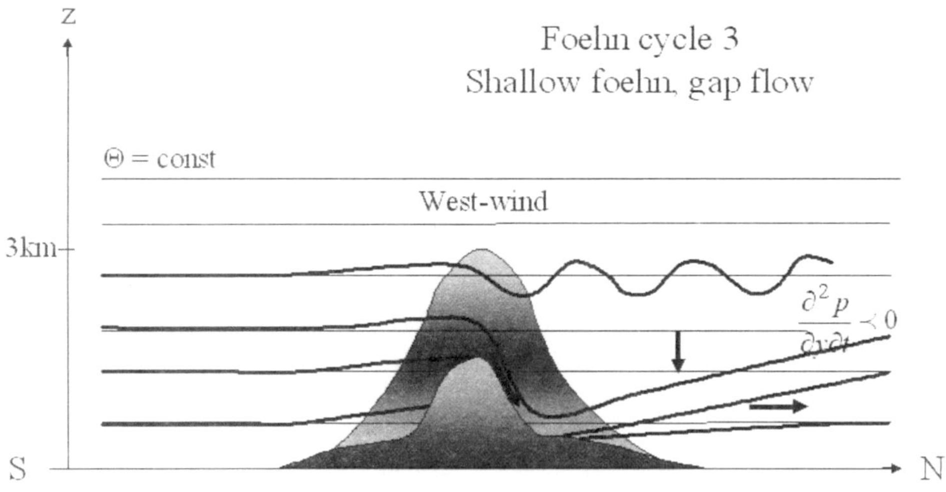

Figure 10. same as Figure 6 but for phase 3. Bold lines are perturbed isentropes

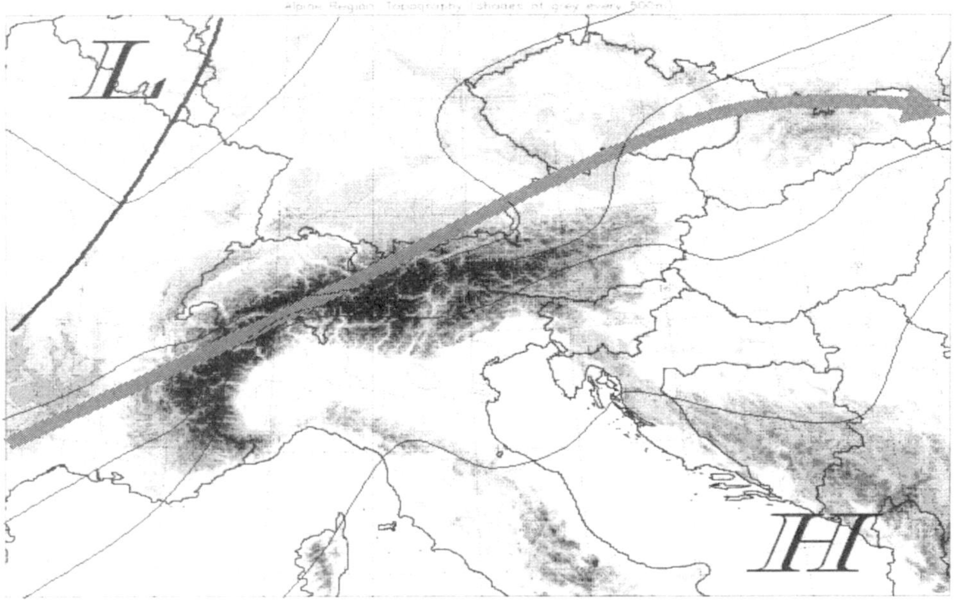

Figure 11. same as Figure 5 but for phase 4. Bold black line represents surface cold front.

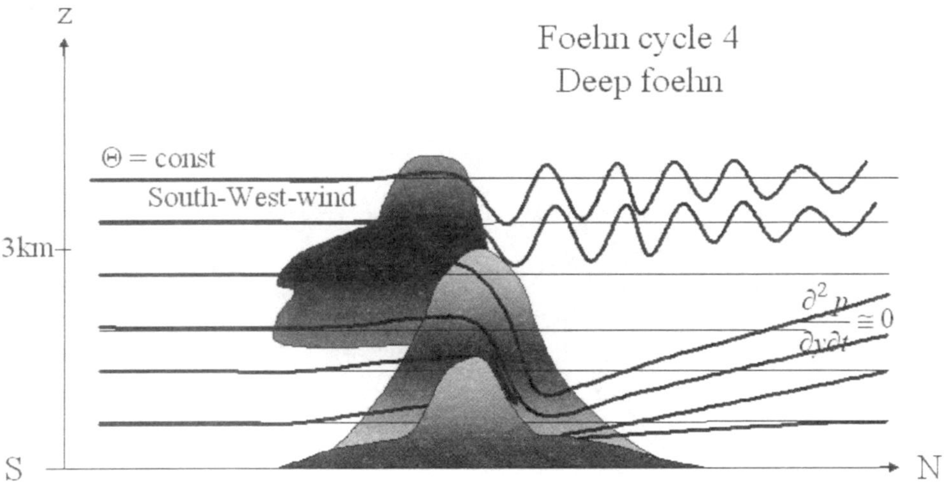

Figure 12. same as Figure 6 but for phase 4. Bold lines are perturbed isentropes

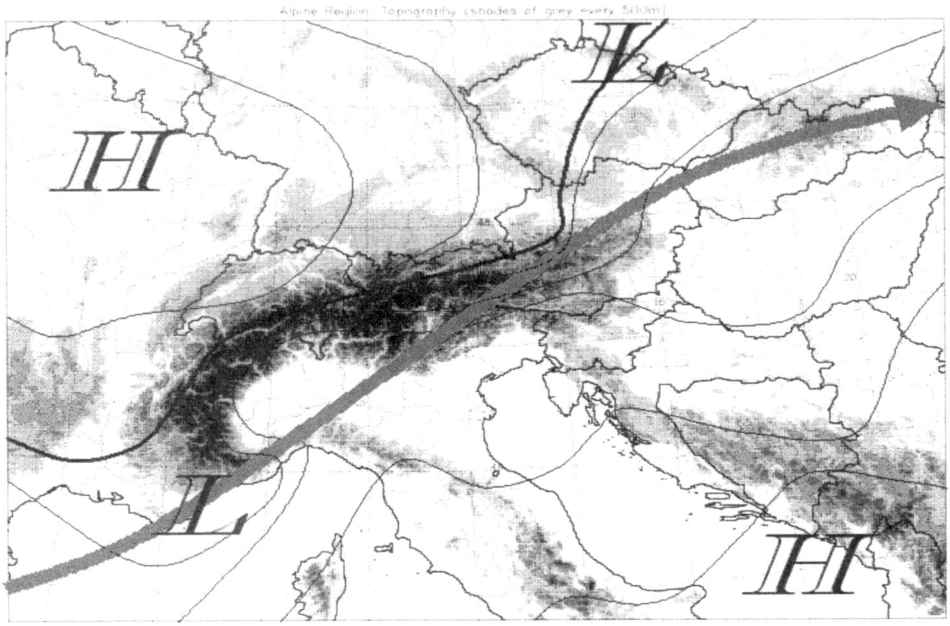

Figure 13. same as Figure 5 but for phase 5. Bold black line represents surface cold front.

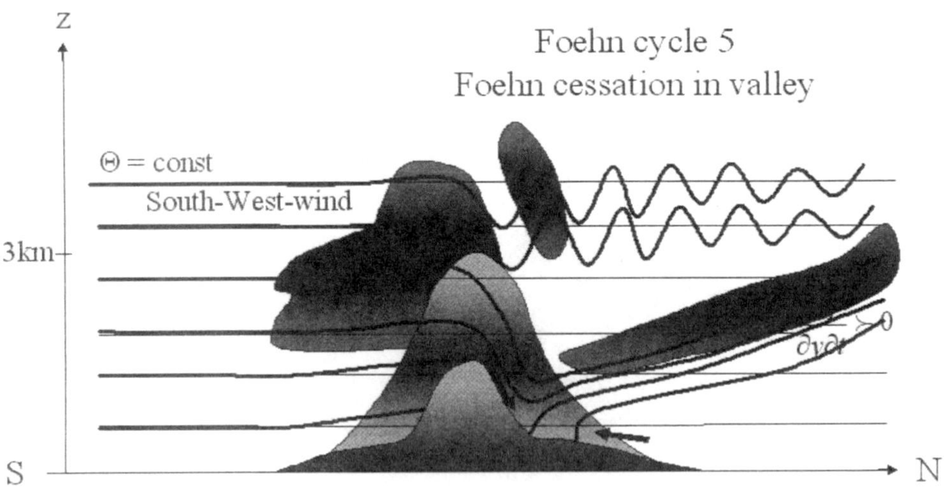

Figure 14. same as Figure 6 but for phase 5. Bold lines are perturbed isentropes.

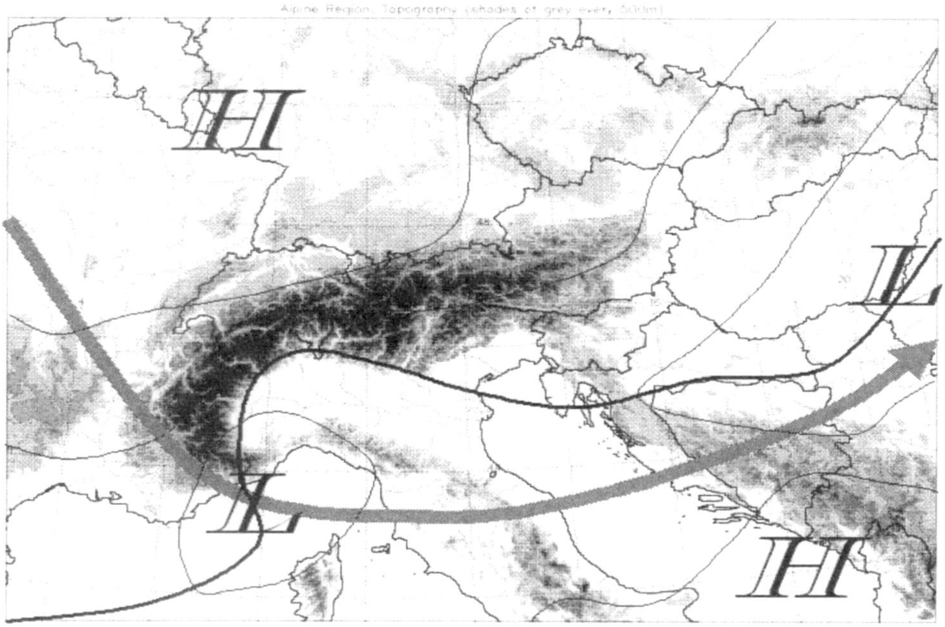

Figure 15. same as Figure 5 but for phase 6. Bold black line represents surface cold front.

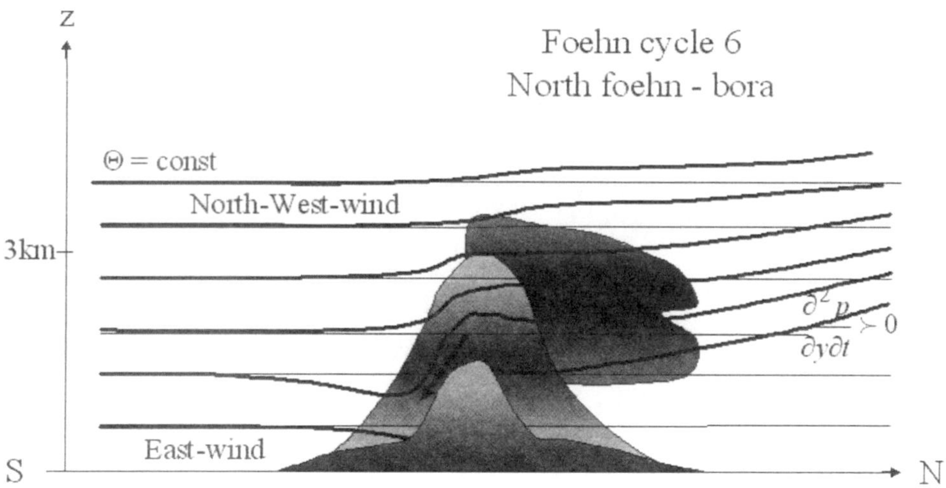

Figure 16. same as Figure 6 but for phase 5. Bold lines are perturbed isentropes.

of the Alps anymore. At crest height the deep foehn has reached its maximum intensity, creating deep lee waves. During the final phase 6 (Figures 15 and 16) the cold air behind the front reaches the crest height of the Alps, so that the low level pressure gradient and wind direction turns its direction and the set in of North foehn, which has altogether bura like characteristics, will be likely. When the baroclinic system has further passed to the East, again a high pressure system will establish over the Alps, we are back to phase 1.

3 Fronts and Orography

In the mid latitudes, fronts are causing most of the precipitation and severe weather. Not only stratiform but also convective precipitation is directly or indirectly linked to fronts. Mountains can modify fronts significantly, what concerns intensity and propagation. This was known since long, see e. g. Figure 17.

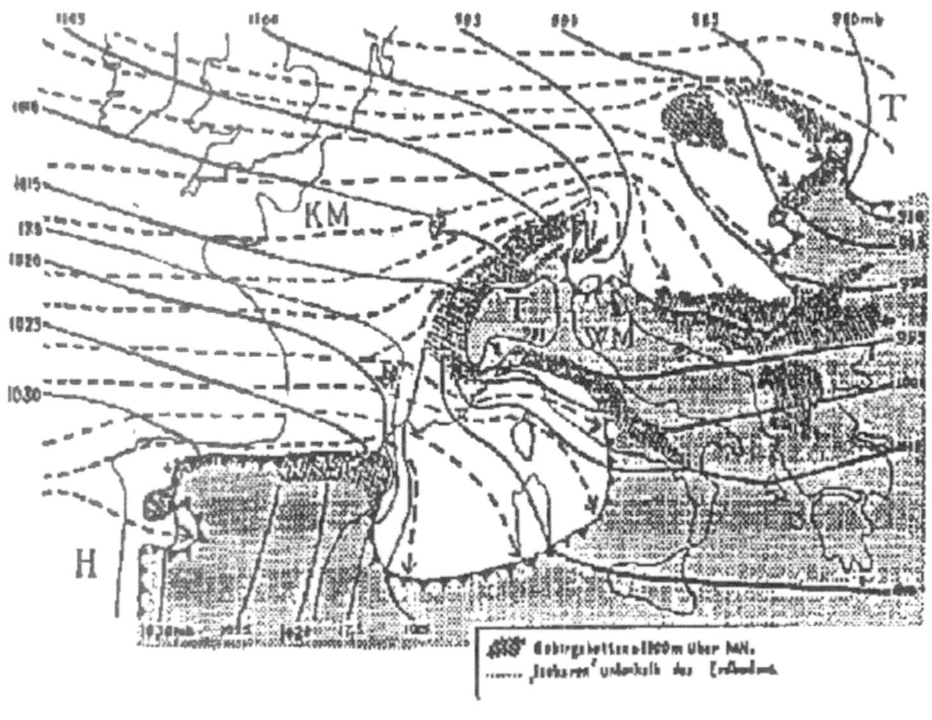

Figure 17. Deformation of a cold front by mountains, taken from Bergeron, 1928. Continuous lined represent surface isobars, broken are surface streamlines

There is an ongoing discussion, if frontal analysis is still meaningful today, when we have mesoscale numerical models, which show the three dimensional structure of the atmosphere with all details. The traditional Norwegian school style of two dimensional large scale frontal analysis might not be the most relevant information of modern synoptic-dynamic meteorology. If we consider the mesoscale and three dimensional frontal analysis, however, it is still an extremely helpful way of condensing the wealth of data and information to the most important atmospheric structures. The frontal analysis can also be automated by determining the extreme values of the second spatial derivatives (see e. g. Steinacker, 1993). An isentropic cross section with three dimensional frontal signatures is shown in Figure 18.

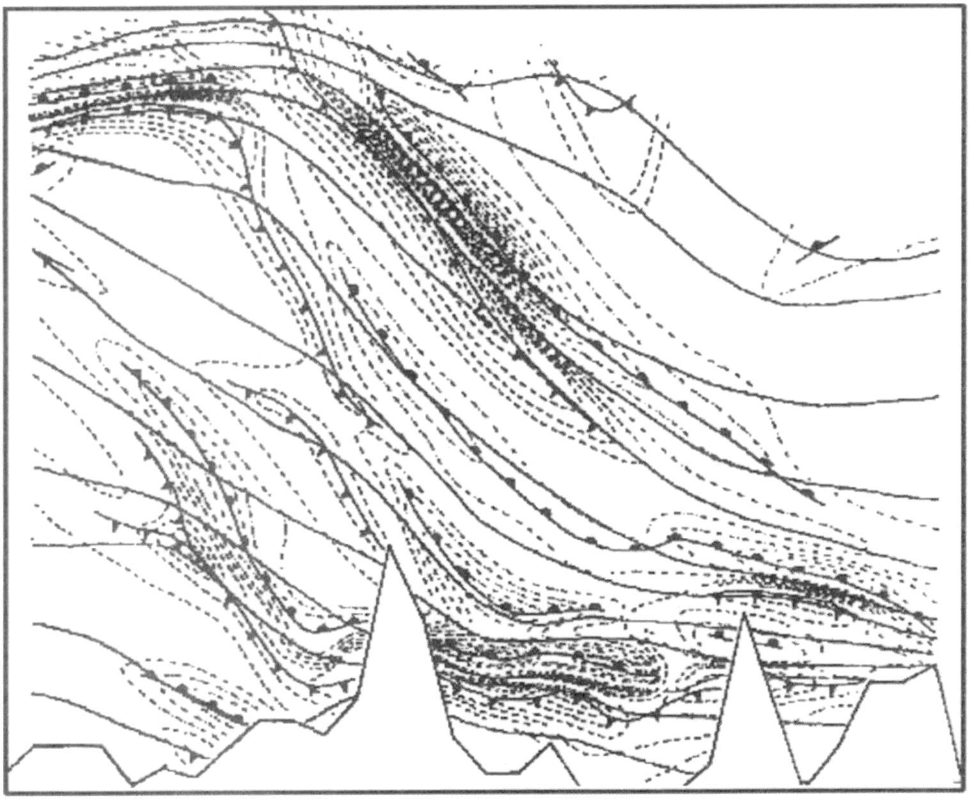

Figure 18. Isentropic cross section between the sea level and the tropopause from Northern Germany (left) over the Alps to Sardegna (right). Continuous are isentropes, dashed are isolines of constant second spatial derivatives. Maxima of these derivatives are denoted by cold air symbols, minima by warm air symbols.

It can be seen, that the broad baroclinic zone consists of several mesoscale frontal features influenced by the mountains especially at low levels.

There is a general tendency that cold fronts impinging mountains are being retarded, e. g. the propagation is reduced. With a three dimensional obstacle the front may propagate faster on both edges of the mountain, leading to a strong deformation (see Figure 19).

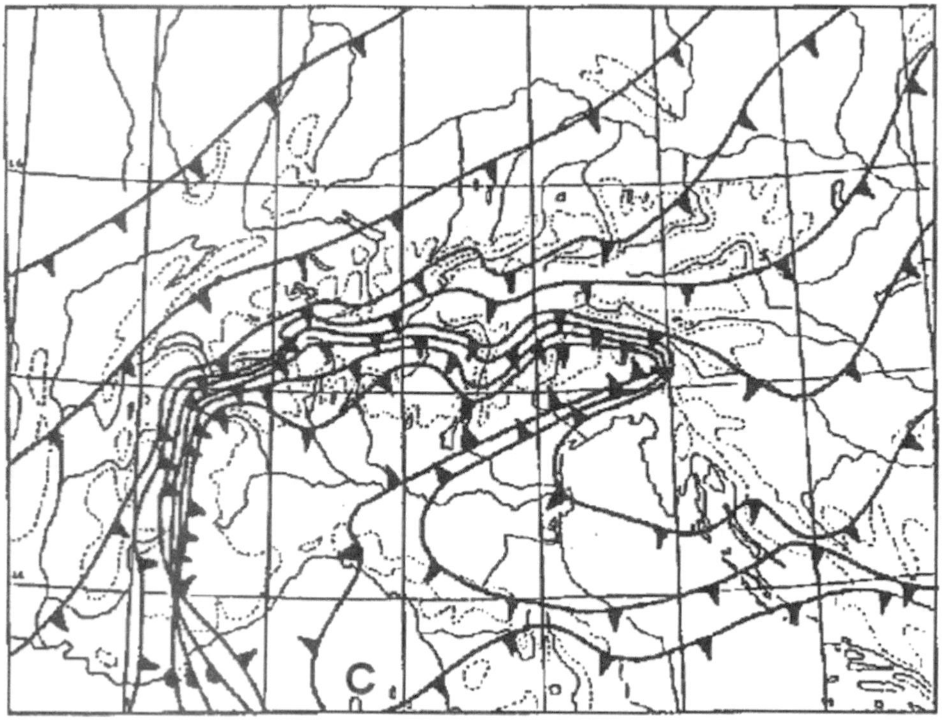

Figure 19. Three-hourly frontal isochrones between 4 March, 00UTC and 5 March, 1982, 18UTC. The limited depth of the cold front did not permit a flow over but rather around the Alps. The cold air reaching the North of Corsica (C) took the long way via the Eastern Alps and the Adriatic Sea.

There are basically two ways, how fronts in combination with topography may enhance convection. The first mechanism is forced lifting on the windward side of mountains. If the pre- or post-frontal air is conditionally unstable, Convection may be triggered for long time over the same location, leading to heavy amounts of precipitation (see e. g. Bougeault, 2001). Mesoscale features of the topography like convex boundaries, may produce some hot spots for convective precipitation, like in the Julian Alps or in the Ticino. Trapped or channelled flows even by minor mountain ranges play an important role for such processes, see Figure 20. An other mechanism which sometimes gives a surprise for inexperienced forecasters is the lee-side triggering of convection. This may be caused by advection of cooler air aloft, ahead of the surface cold front, which is being retarded by the barrier (see Figure 21).

Another phenomenon on the lee side of mountains is the advection of heated air over the mountains towards the forelands. This stabilizes the lower troposphere so that the conditional instability may reach much larger values than normal. This is causing "loaded gun" soundings

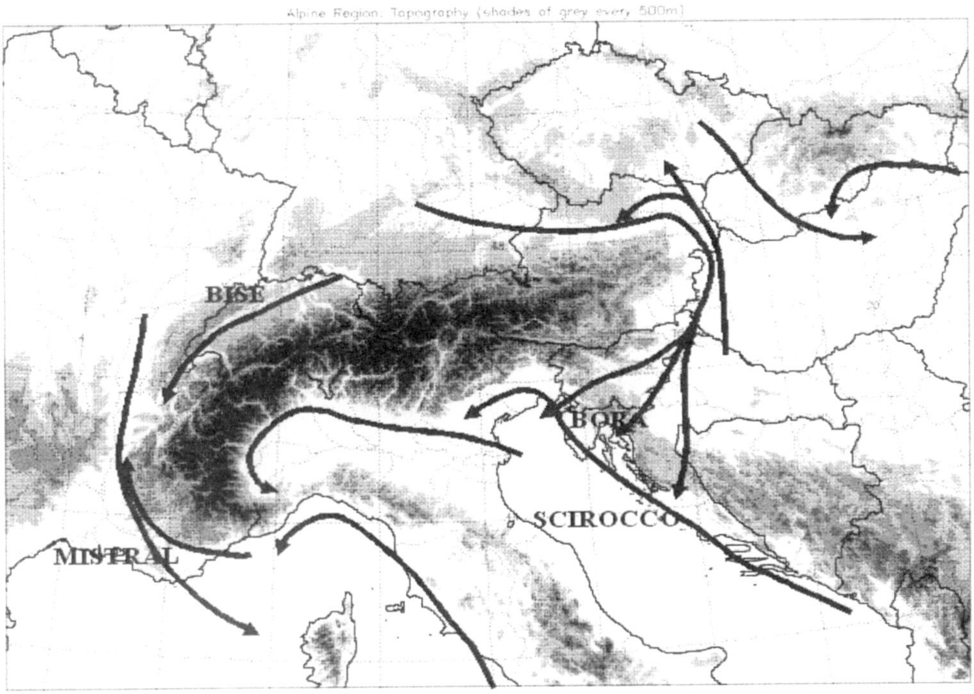

Figure 20. Major trapped and channelled flows around the Alps

in Oklahoma or Kansas due to the warm and dry air intrusion from the highlands of the Rocky Mountains (Bluestein, 1993). Also North of the Alps most severe convective events are taking place during weak to moderate Southwesterly or Southerly flow directions in the middle troposphere, when heated air from the Alps is intruded North-wards, see Figure 22. With stronger Southerly flows the subsidence and drying by foehn is the dominant process, which damps convection. To predict severe convection under such conditions is still a difficult task. Even mesoscale numerical prediction models fail very often due to the extreme sensitivity with respect to small errors in the temperature and wind field.

4 Consequences for Convection

First, the forcing of convection can be caused by lifting of air, flowing towards an obstacle. If the advected air is moist unstable, the triggering of convection does not necessarily occur at the edge of mountains but also in a considerable upstream position, when a blocked low level air is causing the rising motion already at a distance upstream of the mountains. A destabilization of moist stable air (lapse rate less than moist isentropic) by lifting may occur,

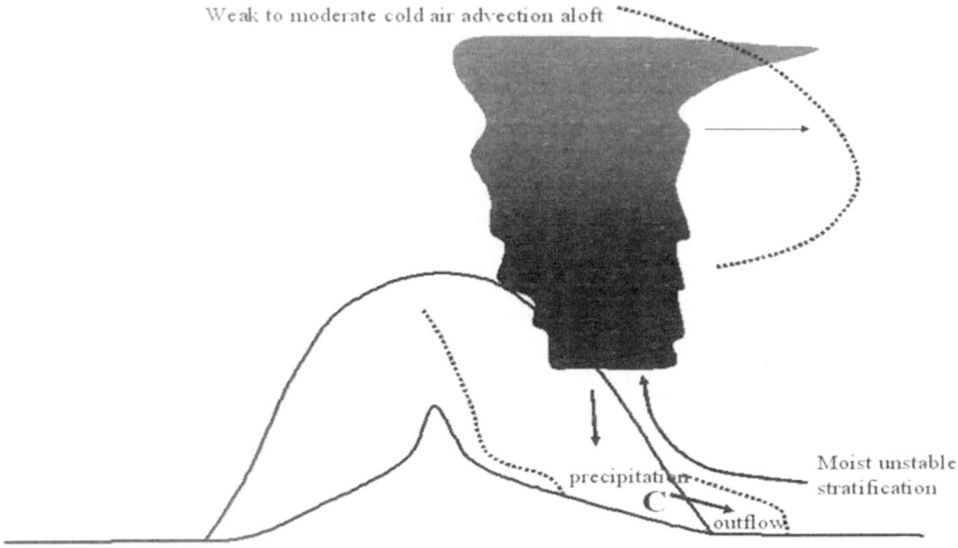

Figure 21. Schematic cross section over a mountain range producing convection on the lee side. Continuous lines represent the lowest and highest profile of topography, the dotted lines represent frontal boundaries.

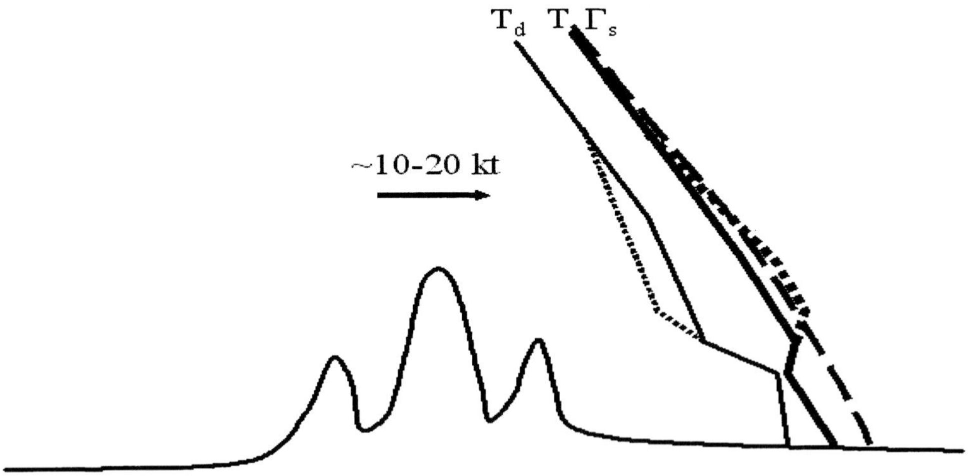

Figure 22. Modification of a vertical temperature (bold) and moisture profile (thin) due to advection of dry and warm air heated by mountains (dotted). One moist isentrope is plotted dashed.

if the equivalent potential temperature is decreasing with height. In such case, the triggering is located close to the rim of the mountains. Such orographically triggered convection may lead to high amounts of precipitation, but rarely to severe convection. The latter is more commonly taking place in a case, when convection is triggered on the lee side of mountains. The mechanism behind lee side convection is either the destabilization through advection of cooler air aloft in conjunction with a front. The most severe convection is triggered, when heated dry air is advected to the lee of mountains, leading to an increase of the convective inhibitions. Then large amounts of CAPE can build up and eventually be released. A further role of topography with respect to severe convection is the generation of horizontal as well as vertical sheared flows, which are necessary for tornadic developments.

References

Barry, R. G. (1992). *Mountain Weather and Climate.* Routledge (UK). 432 pp.

Bergeron T (1928) Über die Dreidimensional Verknüpfende Wetteranalyse, I Teil, *Geophys Publ* 5.

Bluestein H B (1993) *Synoptic-Dynamic Meteorology in Midlatitudes.* Vol II: Observations and Theory of Weather Systems. Oxford Univ. Press. 608pp.

Bougeault, P, P. Binder, A. Buzzi, R. Dirks, R. Houze, J. Kuettner, R.B. Smith, R. Steinacker, H. Volkert (2001) The MAP Special Observing Period. *Bull. Amer. Meteorol. Soc* 82, 433-462.

Fliri, F (1975) *Das Klima der Alpen im Raume von Tirol.* Verlag Wagner, Innsbruckm 454pp.

Hann J (1866): Zur Frage über den Ursprung des Föhns. *Zeitschrift der österreichischen Gesellschaft für Meteorologie* 1, S. 257-263.

Mayr, G.J. and Vergeiner, I. and Gohm, A. and Vergeiner, J. and Mayr, R., 2003: Gap flow - an overview and preliminary results. *Österr. Beitr. Meteor. Geophys.,* **29**, 15-54.

Smith R B (1990) Why can't Stably Stratified Air rise over High Ground? In: *Atmospheric Processes over Complex Terrain* (W. Blumen, ed.), Meteorological Monographs, AMS Boston, 105-107.

Steinacker R (1993) Dynamical Aspects of Frontal Analysis. *Meteor Atmos Phys* 48, 93-103.

Steinacker, R (2006), Alpiner Föhn – eine neue Strophe eines alten Liedes. *Promet: Gebirgsmeteorologie.* Deutscher Wetterdienst, Offenbach, in print.

Thermodynamic Aspects of Topography: the Role of Elevation

Reinhold Steinacker

Institute of Meteorology and Geophysics, University of Vienna, Austria

Abstract. In this document the thermodynamic aspects of topography is being discussed and the consequences for forcing or damping convection. After the discussion of the surface energy budget in dependence of elevation, the generation of thermally driven slope wind circulations is treated. Finally the differential heating of air in valleys due to the volume reduction is derived and its consequences for valley wind systems. Concluding some concepts of convection over complex terrain are discussed.

1 Surface energy budget in dependence of elevation

If we consider the surface energy budget in detail (see Figure 1), we can see that during daytime the short wave incoming as well as the short wave outgoing (reflected) radiation is typically increasing with height. In contrast, the long wave incoming as well as the long wave outgoing radiation is decreasing with height. The total radiation budget, being downward

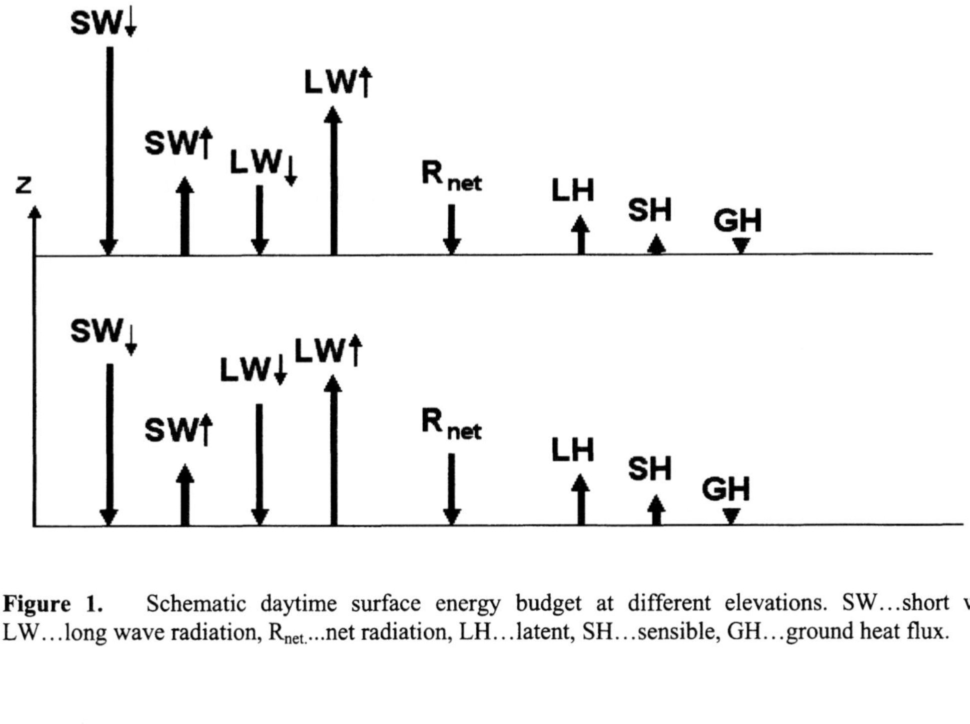

Figure 1. Schematic daytime surface energy budget at different elevations. SW…short wave, LW…long wave radiation, R_{net}...net radiation, LH…latent, SH…sensible, GH…ground heat flux.

oriented during daytime, is decreasing with height. This means that there is less energy available at elevated surfaces for heat fluxes from the surface into the atmosphere and into the ground. Whereas the ground heat flux will not generally change with height, the sensible and especially the latent heat flüxes will typically decrease slightly with height.

During night time the radiation energy loss is typically increasing with height especially due to the decreasing magnitude of atmospheric radiation. The sensible heat flux is generally more important during night as compared to the latent heat flux (see Figure 2)

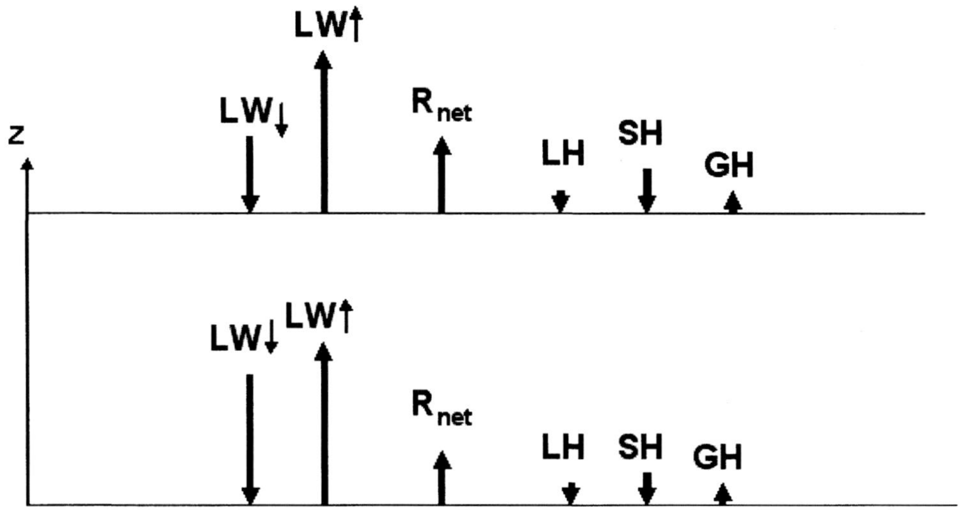

Figure 2. Same as Figure 1, but for night time conditions.

It should be kept in mind that variations in slope angle, slope azimuth, latitude, season, daytime, vegetation, snow cover (albedo!), etc. have a strong impact on the individual terms of the surface energy balance. Let us assume for simplicity at the moment, however, that the energy available for the turbulent energy transport (sensible heat flux) from the ground into the atmosphere or vice versa decreases slightly, proportional to the decrease of air density, with elevation. Then we can compare the temperature profiles of the atmosphere during daytime and night time (see Figure 3).

If we start with a moderately stably stratified atmosphere, e.g. the standard atmosphere conditions, the heating during daytime will lead to a mixed layer with a constant potential temperature. The depth of this mixed layer under the above mentioned conditions will be the same, for a plain at sea level and at an elevated plain. Due to the stably stratified atmosphere the potential temperature will be higher at the elevated plain. During night time a cold surface layer (inversion) will form, again with the same depth at sea level as at an elevated plain. On neither level a dynamical response will be observed if the low and elevated plains are

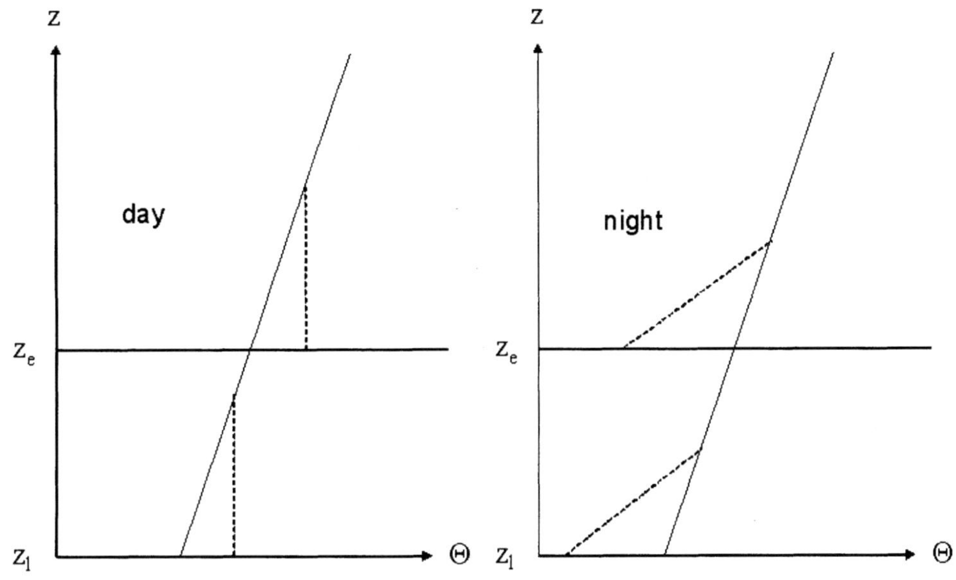

Figure 3. Schematic vertical potential temperature profile after daytime heating (dashed, left side) and night time cooling (dashed, right side) at a plain low level z_l and elevated plain level z_e.

infinitely extending. If, however, a plain at sea level is adjacent to an elevated plain, a baroclinic zone is created, which forces a circulation (see Figures 4 and 5).

2 Slope winds

In reality we will rarely find a stepwise topographic configuration like in Figures 4 and 5, but rather inclined slopes. If we heat or cool slopes we get a temperature perturbation along the slopes, positive during the day, negative, during the night (see Figures 6 and 7). The slope winds can be ideally described like any other thermally driven circulation by the vorticity equation

$$\frac{d\mathbf{Z}}{dt} + \mathbf{Z}\nabla \bullet \mathbf{v} - \mathbf{Z} \bullet \nabla \mathbf{v} = \nabla p \times \nabla \alpha + \nabla \times \mathbf{F} \tag{2.1}$$

with **Z** being the 3D vorticity vector, **v** the wind vector, **F** the friction vector and p and α the pressure and specific volume. In the simplest 1D case of a sufficient long and broad slope the

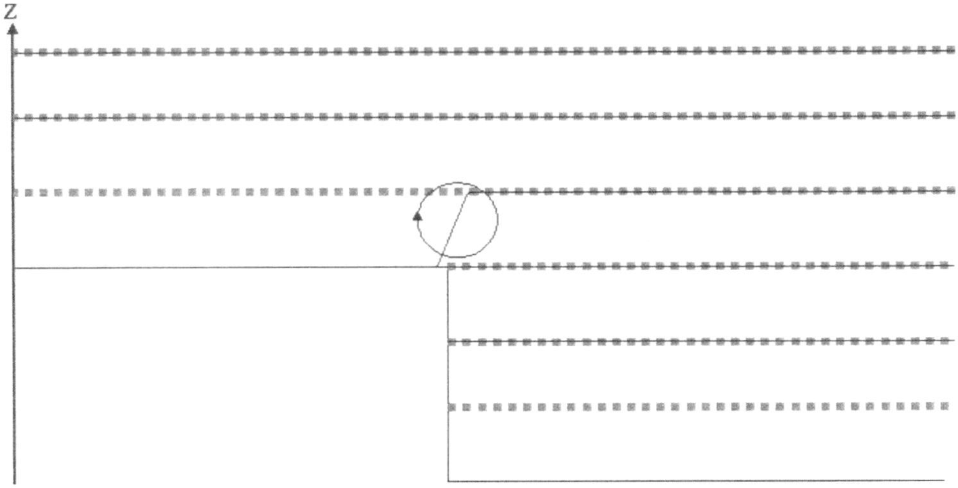

Figure 4. Schematic potential temperature distribution over two plains at different elevations (block mountain). The dashed grey lines are isentropes before heating (stably stratified, i. e. increasing with height) and continuous grey lines after heating, i. e. the lowest isentropes above terrain are removed. The direct thermal circulation response is indicated by the circle with arrow.

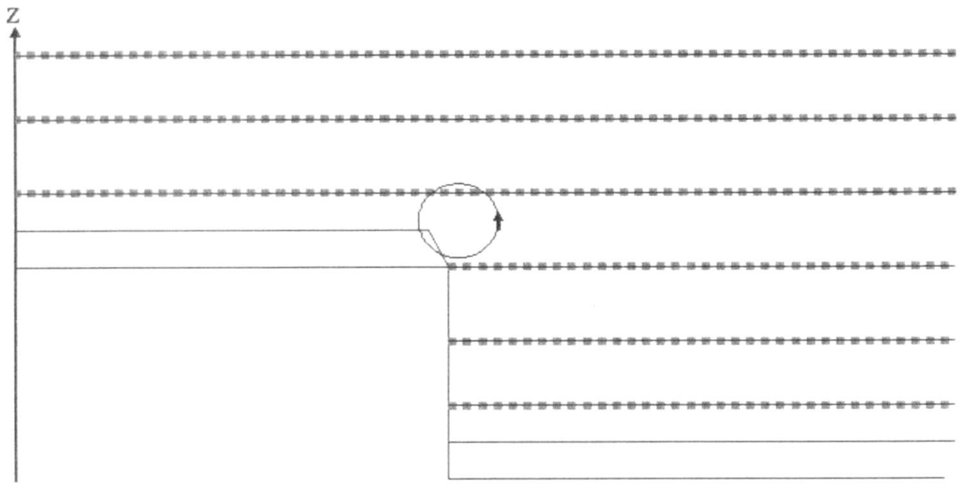

Figure 5. Same as fig. 4 but for night time cooling. An extra isentrope is occurring above the ground

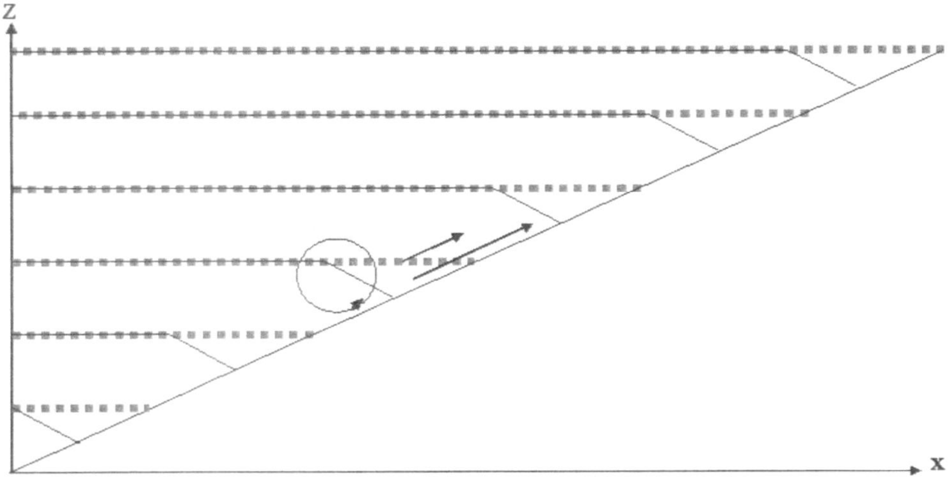

Figure 6. Schematic potential temperature distribution before (dashed) and after heating (continuous) a slope. The thermally driven circulation leads to an upslope flow.

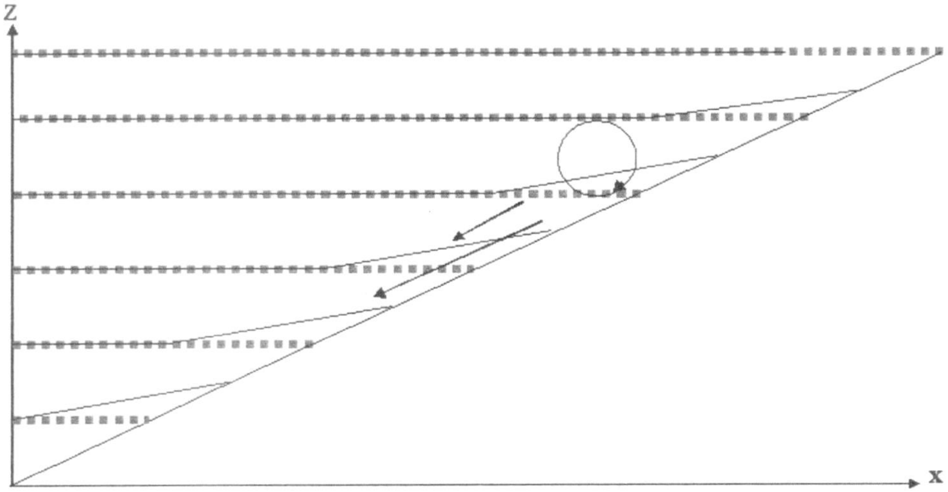

Figure 7. Same as figure 6 but for night time conditions with cooling of the slope layer.

second (divergence) and third (tilting) term of equation 2.1 are vanishing, the same is true for the advection part of the first term. Hence equation 2.1 becomes

$$\frac{\partial \mathbf{Z}}{\partial t} \cong \nabla p \times \nabla \alpha + \nabla \times \mathbf{F} \cdot \tag{2.2}$$

Using the equation of state and the hydrostatic equation the solenoidal term (first right hand side expression of equation 2.1 and 2.2 may be written for a slope in x-direction (figures 6 and 7) with good approximation as

$$\begin{aligned} \nabla p \times \nabla \alpha = \nabla p \times \nabla \frac{RT}{p} = \frac{R}{p} \nabla p \times \nabla T = \frac{R}{p}\left(\frac{\partial p}{\partial z}\frac{\partial T}{\partial x} - \frac{\partial p}{\partial x}\frac{\partial T}{\partial z}\right)\mathbf{j} \cong \\ \cong \frac{R}{p}\frac{\partial p}{\partial z}\frac{\partial T}{\partial x}\mathbf{j} = -\frac{g}{T}\frac{\partial T}{\partial x}\mathbf{j} \cong -\frac{g}{\Theta}\frac{\partial \Theta}{\partial x}\mathbf{j} \end{aligned}, \tag{2.3}$$

where **j** represents the unit vector in y-direction, *R* the gas constant, g the gravity acceleration, *T* and Θ the temperature and potential temperature respectively. The approximation is based on the assumption that the horizontal and vertical temperature gradient is on the same order of magnitude, whereas the vertical pressure gradient is several orders of magnitude larger than the horizontal pressure gradient. The friction term may be formulated in a simple way as linearly depending on the vorticity

$$\nabla \times \mathbf{F} = -k\mathbf{Z} \tag{2.4}$$

Equation 2.2 then reads

$$\frac{\partial Z_y}{\partial t} \cong -\frac{g}{\Theta}\frac{\partial \Theta}{\partial x} - kZ_y \tag{2.5}$$

2.1 A simple slope wind model

To set up a simple model of slope winds, where *u* and w represent the wind component in x- and z direction respectively, we can furthermore formulate

$$\frac{d}{dt}\left(\frac{g}{\Theta}\frac{\partial\Theta}{\partial x}\right)=\frac{\partial}{\partial t}\left(\frac{g}{\Theta}\frac{\partial\Theta}{\partial x}\right)+u\frac{\partial}{\partial x}\left(\frac{g}{\Theta}\frac{\partial\Theta}{\partial x}\right)+w\frac{\partial}{\partial z}\left(\frac{g}{\Theta}\frac{\partial\Theta}{\partial x}\right)=$$
$$=\frac{\partial}{\partial x}\left(\frac{g}{\Theta}\frac{\partial\Theta}{\partial t}\right)+\frac{\partial}{\partial x}\left(\frac{g}{\Theta}u\frac{\partial\Theta}{\partial x}\right)+\frac{\partial}{\partial x}\left(\frac{g}{\Theta}w\frac{\partial\Theta}{\partial z}\right)-\frac{\partial u}{\partial x}\left(\frac{g}{\Theta}\frac{\partial\Theta}{\partial x}\right)-\frac{\partial w}{\partial x}\left(\frac{g}{\Theta}\frac{\partial\Theta}{\partial z}\right)= \quad (2.6)$$
$$=\frac{\partial}{\partial x}\left(\frac{g}{\Theta}\frac{d\Theta}{dt}\right)-\frac{\partial u}{\partial x}\left(\frac{g}{\Theta}\frac{\partial\Theta}{\partial x}\right)-\frac{\partial w}{\partial x}\left(\frac{g}{\Theta}\frac{\partial\Theta}{\partial z}\right)$$

If we introduce a natural coordinate system with s pointing parallel upward the slope with the inclination α and n pointing normal to the slope into the atmosphere (see Figure 8) we can formulate for the slope wind layer:

$$u = V\cos\alpha; \quad w = V\sin\alpha \quad (2.7)$$

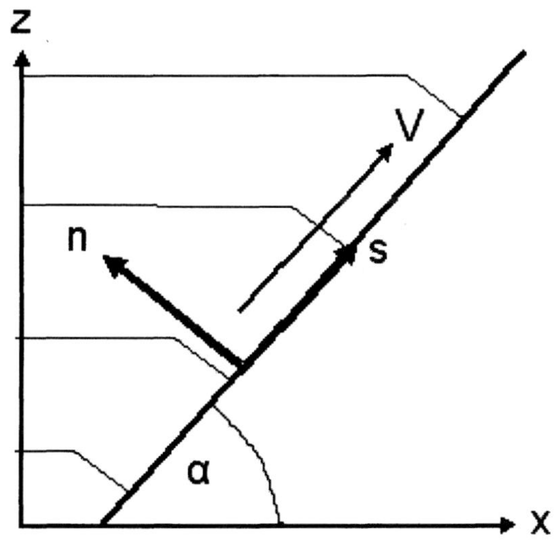

Figure 8. Natural coordinate system with respect to the slope wind. Grey lines represent isentropes.

and with

$$Z_y = \frac{\partial u}{\partial z} - \frac{\partial w}{\partial x} = \frac{\partial V}{\partial n} \quad (2.8)$$

and if we furthermore denote the basic state potential temperature by Θ_0 and the diabatic temperature perturbation by Θ', we can write

$$\begin{aligned}
\frac{\partial u}{\partial x} &= \frac{\partial V}{\partial x}\cos\alpha = \frac{\partial V}{\partial n}\frac{\partial n}{\partial x}\cos\alpha = -Z_y \cos\alpha \sin\alpha \\
\frac{\partial w}{\partial x} &= \frac{\partial V}{\partial x}\sin\alpha = \frac{\partial V}{\partial n}\frac{\partial n}{\partial x}\sin\alpha = -Z_y \sin^2\alpha \\
\frac{\partial \Theta}{\partial x} &= \frac{\partial \Theta'}{\partial n}\frac{\partial n}{\partial x} = -\frac{\partial \Theta'}{\partial n}\sin\alpha \\
\frac{\partial \Theta}{\partial z} &= \frac{\partial \Theta_0}{\partial z} + \frac{\partial \Theta'}{\partial n}\frac{\partial n}{\partial z} = \frac{\partial \Theta_0}{\partial z} + \frac{\partial \Theta'}{\partial n}\cos\alpha
\end{aligned} \tag{2.9}$$

Hence eq. (2.6) may be written as

$$\frac{d}{dt}\left(\frac{g}{\Theta}\frac{\partial \Theta}{\partial x}\right) = \frac{\partial}{\partial x}\left(\frac{g}{\Theta}\frac{d\Theta}{dt}\right) + Z_y \sin^2\alpha\left(\frac{g}{\Theta_0}\frac{\partial \Theta_0}{\partial z}\right) = DH + Z_y \sin^2\alpha N^2 \tag{2.10}$$

where *DH* denotes the differential heating (or cooling).

For our simple slope wind model we have now - under the assumption of zero advection of the solenoidal term, which is true for an infinite slope and a constant static stability - two equations which interrelate the change of vorticity to the change of the baroclinicity:

$$\begin{aligned}
\frac{\partial Z_y}{\partial t} &= -S_y - kZ_y \\
\frac{\partial S_y}{\partial t} &= DH + N^2 \sin^2\alpha Z_y
\end{aligned} \tag{2.11}$$

For a steady state solution, the tendencies of (2.11) must vanish so that

$$Z_y = -\frac{DH}{N^2 \sin^2\alpha} = -\frac{S_y}{k} \tag{2.12}$$

which means the larger the differential heating or cooling, the larger the slope wind will blow and the larger the (background) static stability the weaker the slope wind will blow. It is interesting that due to this simple model it comes out that for a slope with less inclination the slope wind will blow stronger in the equilibrium state than for a steep slope. Furthermore the slope wind will be the stronger the larger the (steady state) solenoidal term is and the less the frictional force is acting.

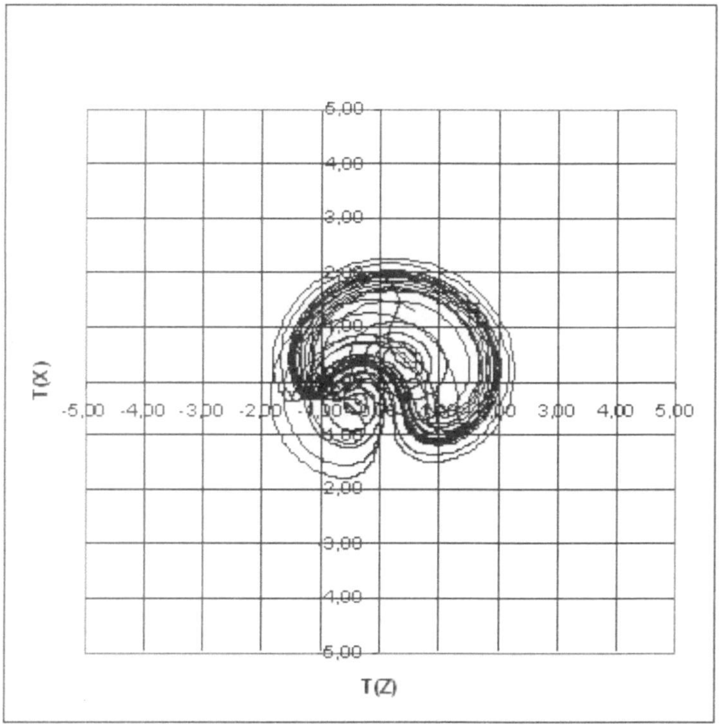

Figure 9. Trajectory in the phase diagram for the slope wind solution of equation 2.11. T(X) represents the horizontal temperature difference, T(Z) the vertical temperature difference between the slope wind layer and the "free" valley atmosphere.

In contrast to more traditional slope wind models (see e. g. Prandtl, 1942, Vergeiner, 1982 or Egger, 1990), we find under certain parameter settings – low frictional force and relatively steep slopes - a chaotic solution (see Figure 9), when computing the numerical solution of the two equations 2.11.

The chaotic solution of equations 2.11 means that slope winds under stably stratified conditions, when the slope angle is large enough and the friction is small enough, are pulsating irregularly. This is observed during the day as well as during the night, when "air avalanches" are commonly observed (Kuettner, 1949).

3 The topographic amplification factor

So far we have only considered the slope wind layer. The slope wind, however, must have an influence on the adjacent air due to mass continuity. Let us assume a valley of width $2D_V$ and a horizontal width of the slope wind layer of D_S, see Figure 10. Then, due to mass continuity, the vertical flow in both slope wind layers must be compensated by an opposite vertical motion in the free valley atmosphere, under the assumption of zero horizontal flow divergence.

$$2w_S(z)D_S = -w_V(z)\left(D_V(z) - 2D_S\right) \tag{3.1}$$

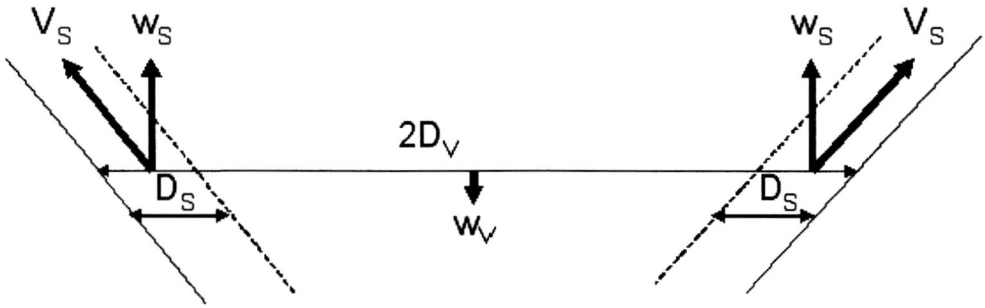

Fig. 10 Schematic view to show the effect of the mass continuity in a valley with respect to vertical motions.

This means that during daytime, when the slopes become heated by solar radiation, the upslope flow produces a subsiding motion in the free valley atmosphere and during the night, when cooling of the slopes is effective with down slope winds, there follows an upward vertical motion in the free valley atmosphere. Although these compensating vertical velocities are much smaller as compared to the slope wind layer, where the typical vertical wind component is on the order of a few m/s, they are thermodynamically important (see Vergeiner, 1982). If the valley width is e. g. 100 times larger than the valley wind layers on both sides, the magnitude of the vertical motion is on the order of a few cm/s. With a moderate stably stratified valley atmosphere the adiabatic warming is then on the order of 1K/h. This means that the warming (cooling) of the slope wind layer is transferred to the whole valley atmosphere by this compensating motion.

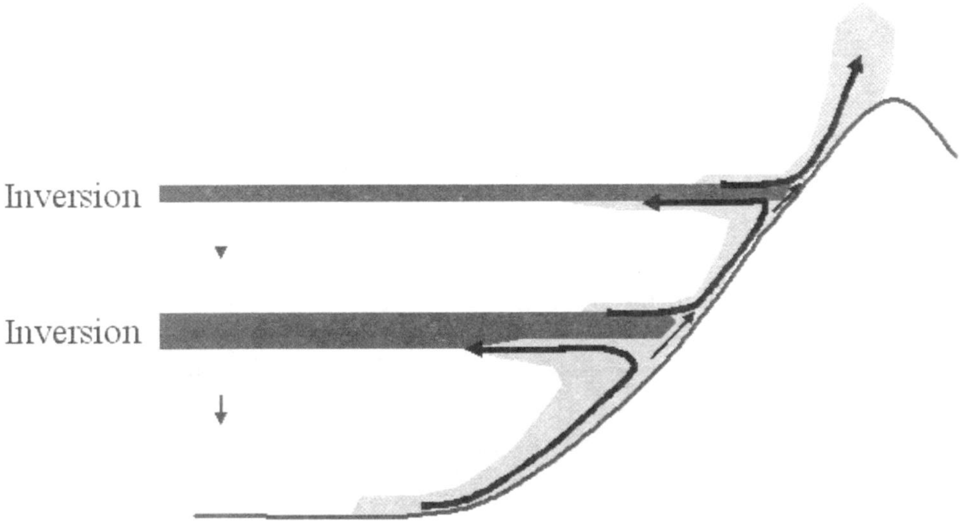

Figure 11. Schematic view of the slope wind (light shading) when two inversions (dark shading) are present, i. e. varying stratification with height. According to equation 2.12 at the elevation of the inversions the slope wind must be much weaker than in the less stable layers, so that more or less closed circulation wheels are generated between inversions.

If in a valley the stratification is varying with height, which is commonly the case, the vertical motion is also varying, leading to partly closed circulation wheels, see Figure 11. This behaviour means that inversion layers may stay for a considerable time in valleys, despite the heating of the whole valley. This behaviour can be observed by haze layers (see Figure 12), which stay often the whole day in the cool season, despite considerable daytime warming.
The differential heating over complex terrain has important consequences. The warming or cooling of a column of air in a valley may be significantly enhanced by a factor of 2 or more. On a climatological average the diurnal temperature amplitude of the Inn valley is larger by the factor of 2.2 as compared to the column of same depth over the Alpine forelands (Wagner, 1932). This has also significant consequences with regard to convection. To quantify this difference we need to know the accurate area height distribution of a valley segment (Steinacker, 1982). This allows us to determine the so called topographic amplification factor TAF (Whiteman, 2000):

$$TAF = \frac{V}{V_{red}} = \frac{A\left(z_{crest} - z_{valley\ floor}\right)}{A\left(z_{crest} - z_{mean}\right)} \quad ; z_{mean} = \frac{1}{A}\sum_i z_i \delta A_i \tag{3.1}$$

Figure 12. Photograph of a multilayer inversion, visible by haze in an Alpine valley, which stays often the whole day, despite of diabatic heating of the atmosphere (photograph by R. Steinacker).

where A is the considered area of a valley segment, A_i is the area above ground at level i and δA_i is the area increment between A_{i+1} and A_i (see Figure 13). If we want to use the TAF for the quantitative derivation of the differential heating we should rather use the mass-specific $\mathrm{TAF_M}$:

$$TAF_M = \frac{\left(p(z_{valley\ bottom}) - p(z_{ridge\ top})\right)}{\left(p(z_{slope})_{mean} - p(z_{ridge\ top})\right)} \tag{3.2}$$

Depending on the area height distribution or hypsographic curve (see Figure 14), the TAF will mostly lie between 1 and 2, occasionally even a bit higher.

The TAF, as outlined above, represents the ratio between the air volume in a valley compared to the air volume over the plains with respect to the same area A. The mass specific $\mathrm{TAF_M}$ represents the ration between the air mass in a valley compared to the Mass of air over the plains with respect to the same area A. We have seen from previous considerations that under statically stable conditions the energy input/loss on the slope is transmitted to the whole

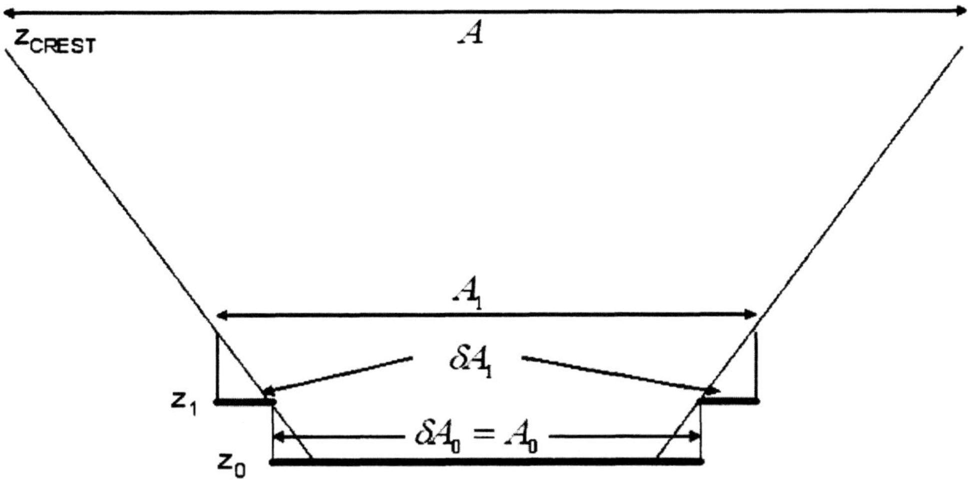

Figure 13. Schematic figure with finite vertical layers to derive the topographic amplification factor TAF

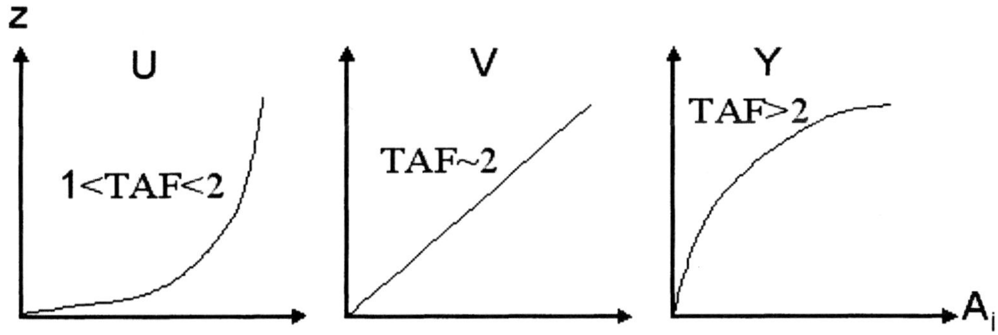

Figure 14. Prototypes of area-height distributions. U-type left, V-type center and Y-type right.

valley atmosphere through compensating vertical motions layer by layer. Therefore it is important to reconsider the idea of the *TAF* towards a differential *TAF (DTAF)*, layer by layer.

$$DTAF_M(i) = \frac{\left(p(z_{layer\ bottom}) - p(z_{layer\ top})\right)_i}{\left(p(z_{mean}) - p(z_{layer\ top})\right)_i} \frac{\delta A_i}{A_i} \tag{3.3}$$

The expression z_{mean} is the area mean elevation of the lower boundary of the air volume within the corresponding layer i. The temporal change of the temperature in each layer of the atmosphere can be calculated with the aid of this *DTAF* if the sensible heat flux SH_i is known:

$$(\Delta T)_i = \frac{SH_i\ \delta A_i}{c_p\ M_i}\Delta t = \frac{g\ SH_i\ \delta A_i\ \Delta t}{c_p\ A_i\left(p(z_{mean}) - p(z_{layer\ top})\right)_i} = \tag{3.4}$$
$$= \frac{g\ SH_i\ \Delta t}{c_p\ (p(z_{layer\ bottom}) - p(z_{layer\ top}))_i} DTAF_M(i)$$

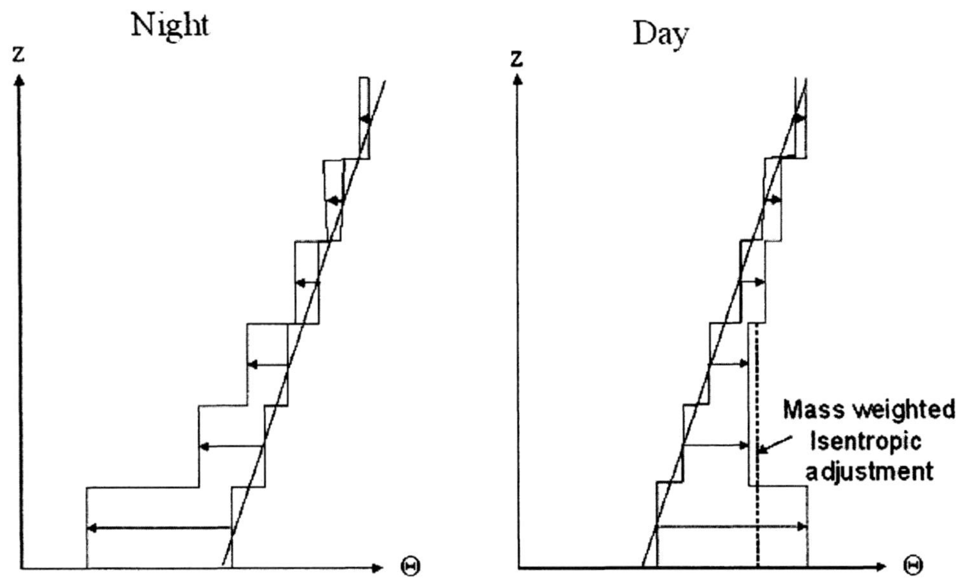

Figure 15. Schematic temporal temperature change in a valley during cooling (left) and heating (right) applying the DTAF to layers of finite depth. The initial state is moderately stably stratified, e. g. according to the standard atmosphere lapse rate.

The schematic application of the temporal temperature change during layer-wise cooling and heating is shown in Figure 15.

Heating the lowest layers individually will soon yield an unrealistic super adiabatic lapse rate. To avoid this, a mass weighted isentropic adjustment has to be carried out. If the finite

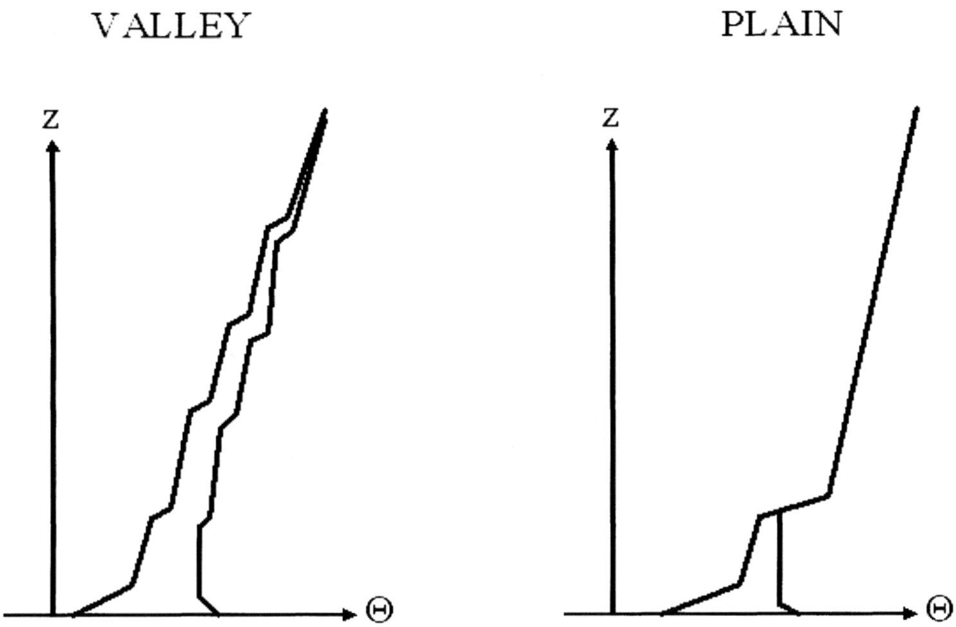

Figure 16. Typical stably stratified vertical profile of the potential temperature in the morning (left curve) in a valley and over a plain. The right profile is after some time of heating.

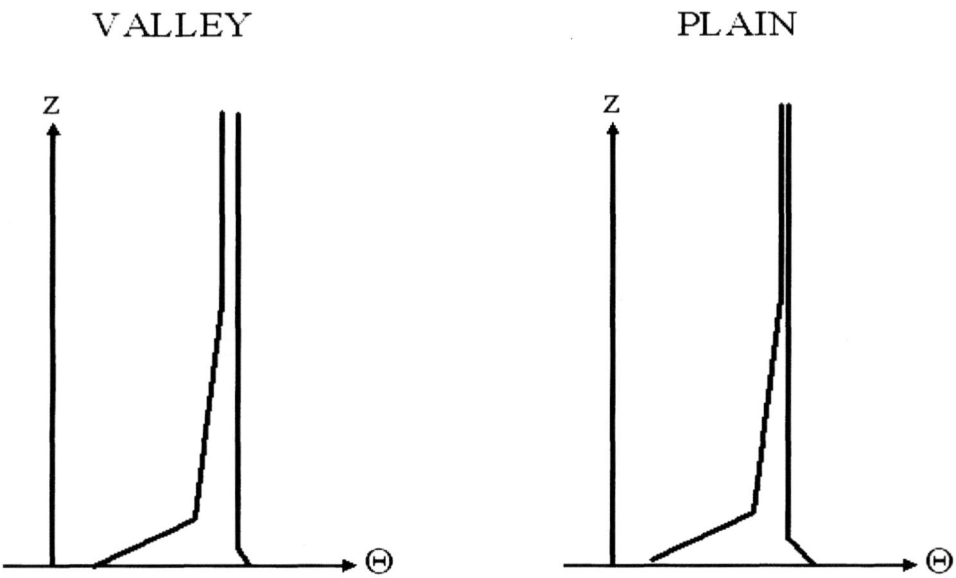

Figure 17. Same as Figure 16 but for an initially weak stable atmosphere.

depth of the layers is kept sufficiently small, we receive vertical temperature profiles for a realistic, i. e. variable initial stratification like in Figures 16 and 17. It can be easily recognized that under stably stratified initial conditions (Figure 16) the ratio of warming or cooling of a valley atmosphere as compared to the atmosphere over flat terrain may exceed a factor of two by far (Steinacker, 1982). If the stratification is less stable (Figure 17), the ratio may be close to unity.

What we can learn from the two Figures 16 and 17 is that the temperature evolution in a valley cannot be described by a constant amplification factor (see also Whiteman et al, 2004b), but the ratio has to be calculated in dependence of the static stability. Furthermore, in a valley we cannot use standard procedures for estimating the daytime evolution of the temperature profile or the cumulus condensation level (CCL) just by given energy input data. Finally we see that a prediction model which does not use a realistic area height distribution will not be able to predict correct temperature profiles and hence convective processes. An example of a temporal evolution of the vertical temperature profile as observed in a sink hole is given in Figure 18. In contrast to a plain the heating does not produce a well mixed surface layer but the heating along the slopes and its transfer layer by layer keeps the stability during the whole warming process high.

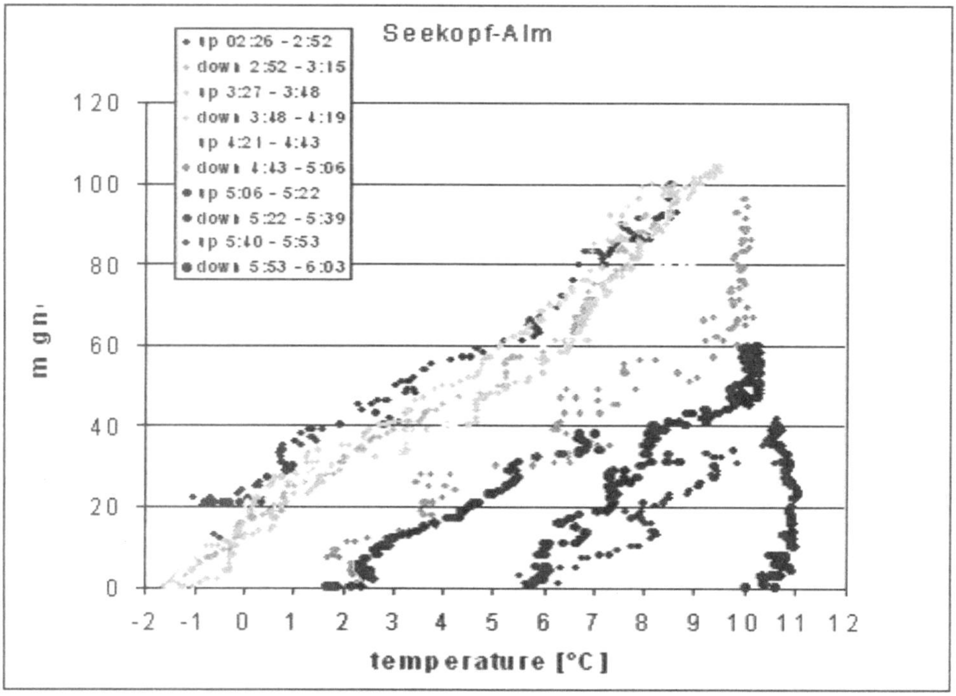

Figure 18. Vertical temperature profiles in a sinkhole during the first 3 hours of the heating period after sunrise (after Whiteman et al, 2004a).

4 Valley winds

The differential heating of the valley atmosphere in contrast to the atmosphere over the plains produces a hydrostatic pressure difference. This is the motor for the valley wind circulation. If we assume during a synoptic weak gradient weather situation the pressure at crest height being the same over the valley and over the plains, the stronger daytime warming or night time cooling of the valley atmosphere produces hydrostatically a daytime low pressure and night

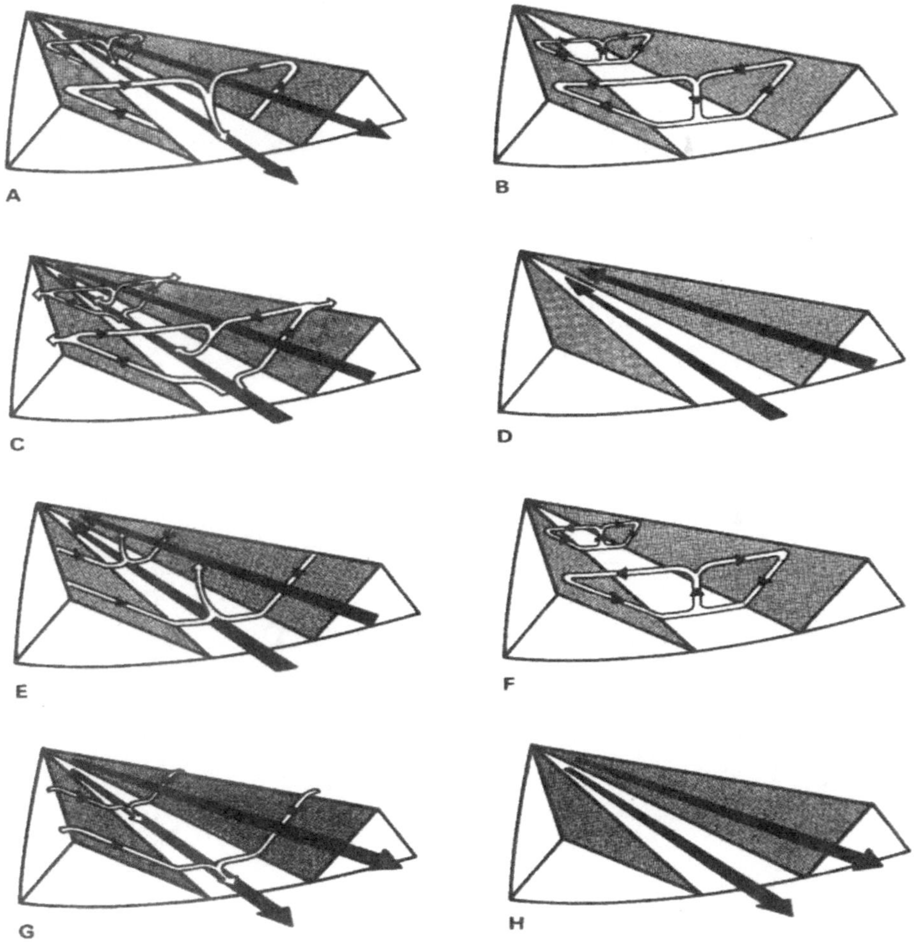

Figure 19. Schematic diurnal evolution of the slope- and valley winds in a valley: A...after sunrise, B...late morning, C...at noon, D...at sunset, E...after sunset, F...late evening, G... at midnight, H...at sunrise, after Defant, 1949.

time high pressure at the valley floor. Observations show that especially during periods with a pronounced stably stratified atmosphere, e. g. in autumn and winter, thermal pressure patterns (heat lows and cold highs) are best pronounced, in agreement with our theoretical considerations. Whereas the slope wind circulation is reacting immediately upon changes of the energy balance of the slope, e. g. the upslope winds start minutes after sunrise and the downslope winds start minutes after sunset, the valley wind circulation is reacting much slower. Although the warming of the valley atmosphere after sunrise is stronger than the warming of the atmosphere over the plains, the lower overall temperature of the valley atmosphere at sunrise needs several hours to reach the same temperature as over the plains. The same is true for sunset time, when the temperature of the valley is considerably higher than over the plains. Therefore the stronger cooling of the valley takes several hours to produce a temperature equilibrium between valley and plains. The schematic diurnal development of slope and valley wind systems is plotted in Figure 19. In reality it should be considered that the valley geometry will not generally lead to the same surface energy budget and sensible heat flux at both sides (slopes) of the valley. Under certain circumstances only one slope may be insolated, which is then causing a non-symmetric circulation, eventually a circulation across the whole valley. It is interesting to note, that, for a vertical air column above a slope, the surface energy budget may become higher than the maximum possible value over flat terrain (see Figure 20).

The net short wave radiation flux hitting a plain unit surface f is given by Lambert's law

$$SW_P = fI(1-A)\cos\delta\,, \tag{3.5}$$

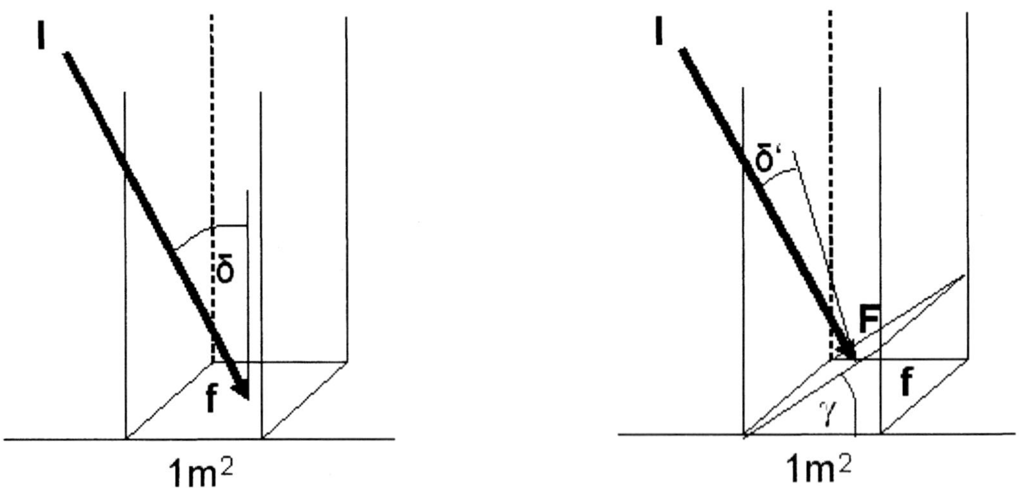

Figure 20. Schematic sketch of an air column over a plain (left) and over a slope (right) absorbing solar radiation.

where I denotes the solar irradiance, i. e. the solar flux density through a unit area normal to the radiation flux vector and A denotes the Albedo. If we consider an inclined slope area F above a horizontal unit area f, the net shortwave radiation hitting F is

$$SW_S = FI(1-A)\cos\delta' = \frac{f}{\cos\gamma} I(1-A)\cos\delta' \ . \tag{3.6}$$

This means that even with a low height angle of the sun a steep slope facing to the sun may receive considerable amounts of energy per horizontal unit area, exceeding the maximum possible value over a plain by far. The redistribution of this energy surplus through the slope wind circulation and the corresponding subsiding vertical motion of the "free valley atmosphere", however, leads to the same total amount of heating of the valley atmosphere as

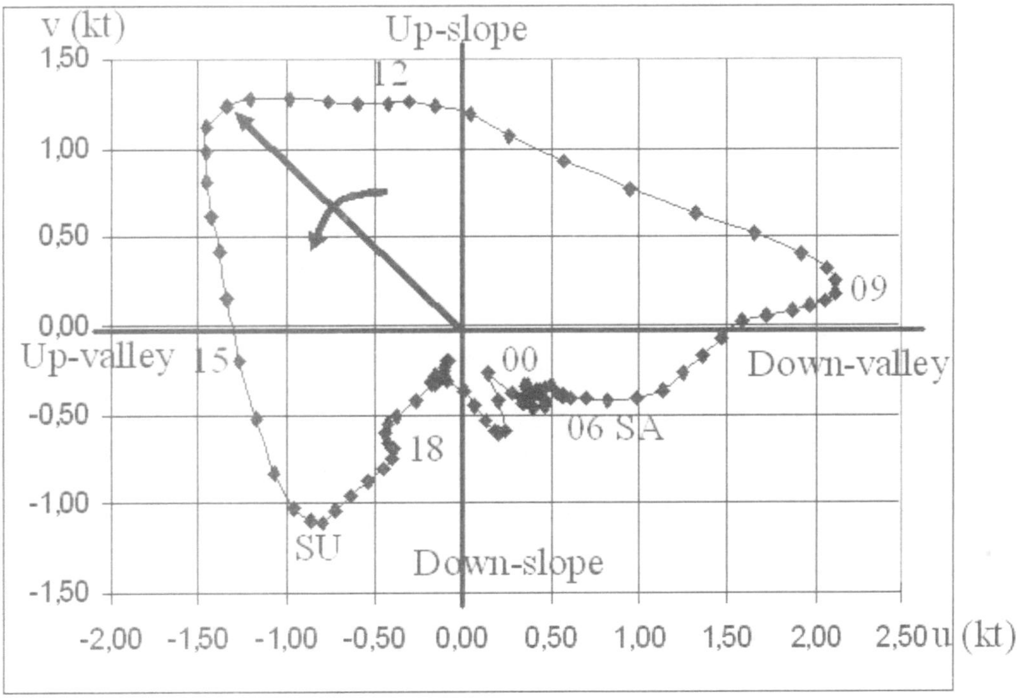

Figure 21. Diurnal variation of the wind at an Alpine slope station (courtesy R. Steinacker) close to Innsbruck. Data are monthly (October 2001) mean values for each 15 minutes (UTC) of the day. SA means astronomical sun rise, SU means astronomical sun set.

in the case of an evenly distributed radiation input - e. g. with a valley axis directed towards or against the Azimuth of the sun.

An observed diurnal variation of the wind on a south facing slope of an eastward draining valley is given in Figure 21. It is nicely shown that the wind varies counter clock wise, which means that the slope wind is reacting much faster to the change of sign of the (slope) surface energy balance than the valley wind. Note that the change of sign of the (slope) surface energy balance does not coincide with the astronomic sun rise or sun set. The difference between mean local time and UTC is approximately 40 minutes. It is interesting furthermore, that the down slope wind speed decreases in the course of the night. This is due to the increasing static stability of the valley atmosphere during the night which leads to a decreasing slope wind even with a constant energy budget (see equation 2.11).

The geometrical reduction of an air volume in a valley or a basin has also further thermodynamical consequences. When computing the convective available potential energy CAPE from a sounding in a valley one has to remember that an air column over a valley segment has a reduced volume up to crest height. Hence the same value of CAPE over a valley and over a plain does not necessarily mean the same.

Very often the moisture in a valley during deep convection is lower than over adjacent plains. This may be also explained by the volume reduction in the valley. The mixing of originally moist air in the valley with dryer air aloft has to be weighted by the respective volumes or masses of air repectively.

5 Diabatic cooling of valleys by precipitation melting

An other common feature is that during precipitation the air in valleys is lower than over the adjacent plains, which implies that e. g. the level of snowfall is lower over mountainous terrain. Also this effect may be easily explained by the reduced air volume in valleys and basins. When solid precipitation is falling below (geometrically) the 0-degree level of the wet bulb temperature, the ice or snow particles start to melt. The melting energy is taken from the sensible and latent energy of the melting layer. The corresponding cooling is destabilizing the atmosphere below the melting layer. This means that depending on the initial stability of the air after some time the whole air in the valley is contributing to provide the energy for further melting. Whereas over the plains the whole column of air below the melting is providing this energy, in valleys or basins only the reduced volume/mass is available. This leads – when the same initial conditions are present, only little advection takes place and with the same amount of precipitation - to a stronger cooling of the valley air as compared to the plains. Fig. 22 shows an example, where this effect was documented clearly by observations.

6 Consequences for convection

If we look for schematic pictures of topographic forcing of convection (e. g. Houze, 1993) we can find the distinction between centered forcing with weak synoptic flows, windward forcing due to dynamical upward motion and leeward forcing due to leeside counter flows (see Figure 23). One determining factor is often neglected, the relative height of topography with respect

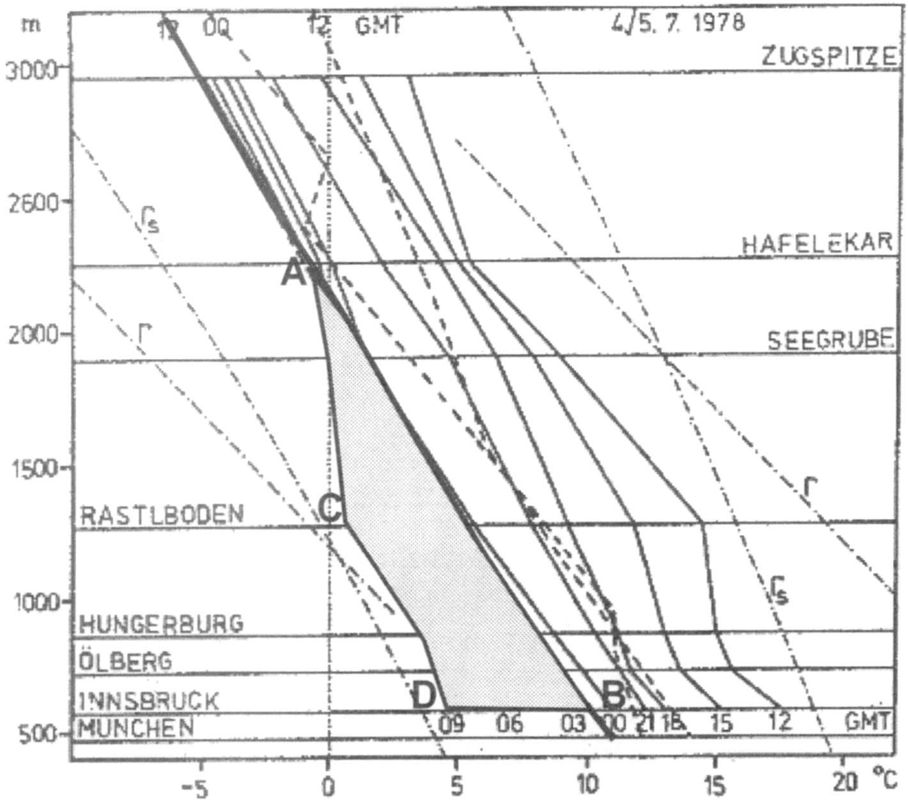

Figure 22. Vertical temperature profiles during a precipitation event in the Inn valley leading to a strong diabatic cooling by melting precipitation. Dash-dotted are dry and moist isentropes. Continuous lines are observed vertical temperature profiles in a valley during a cold front passage at three hourly intervals, see lower numbers. The profile A-B represents the air mass behind the cold front, A-C is the nearly isothermal melting zone and C-D is the layer which has provided the melting energy. The dashed lines are the temperature profiles over the Alpine forelands at 12 hourly intervals, see upper numbers.

to the boundary layer depth of adjacent plains. If the mountains are lower than the typical depth of the boundary layer, mountains are very effective in triggering convection. For the Central European region the boundary layer depth is approximately 1500m. This implies that all mountain ranges of relative height (height above surrounding plains) below 1500m, like the Black Forrest, the Vosges, the Jura mountains are well known hot spots of convection. For mountains of significant higher relative height like the Alps, the convection is rather concentrated at the rims (see Figure 24). The multi range character of the Alps leads to a significant drying of the inner Alpine (valley) atmosphere (see Figure 25) and hence to a

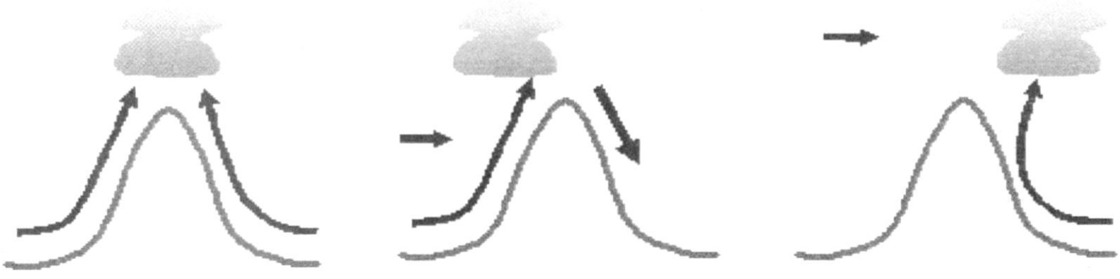

Figure 23. Textbook schemes of orographically induced convection. Centered triggering with weak flow conditions (left), upstream convection due to forced lifting (center) and leeside triggering by counterflows (right).

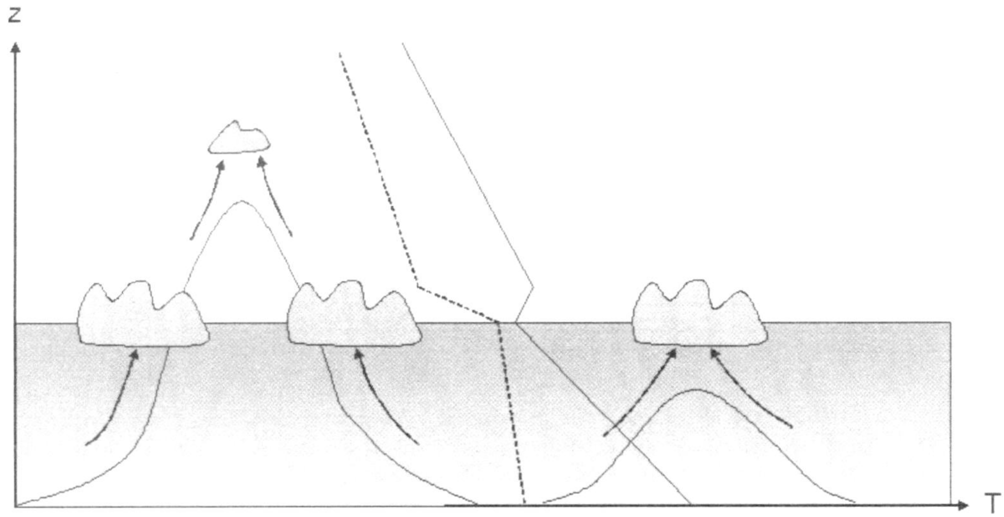

Figure 24. Convection in dependence of boundary layer depth (shaded). High mountains (left) trigger convection at the both rims and lead eventually to a secondary convection over the crest. For low mountains (right) the convection is strongest over the crest. Schematic vertical temperature (continuous) and dew point temperature (dotted) is plotted in the center.

damping of convection. This is clearly evident in climatologies of Alpine thunderstorm frequency, where a pronounced minimum is found over the central Alpine region (see e. g. Fliri, 1975).

The temperature surplus in valleys during the day due to the topographic amplification is obviously compensated by the drying. This leads to a significant increase of the cumulus condensation level (see Figure 26).

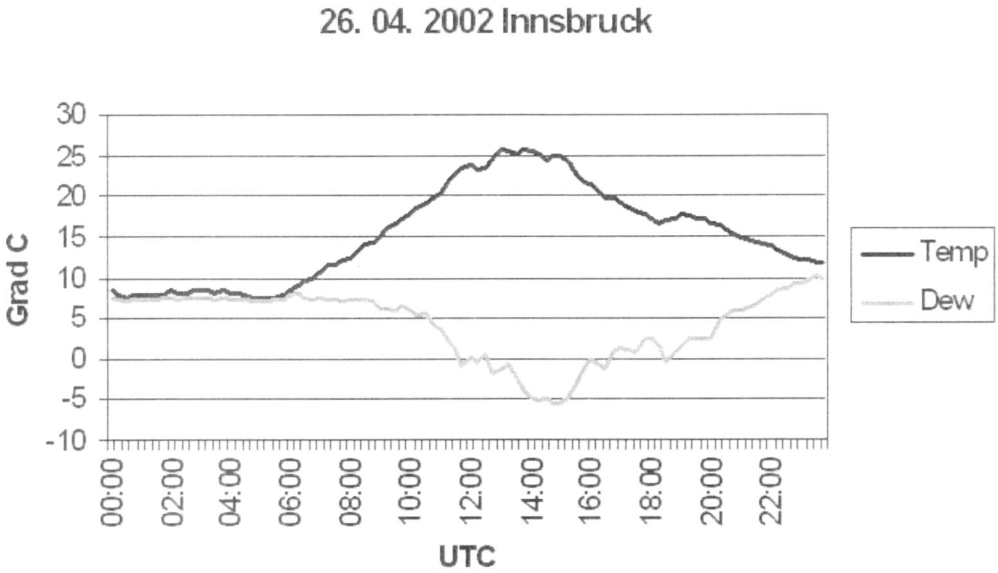

Figure 25. Example of a diurnal temperature and humidity (dew point temperature) profile in a mayor Alpine Valley (Innsbruck). Whereas the temperature rises by nearly 20 degrees, the dew point temperature drops by more than 10 degrees.

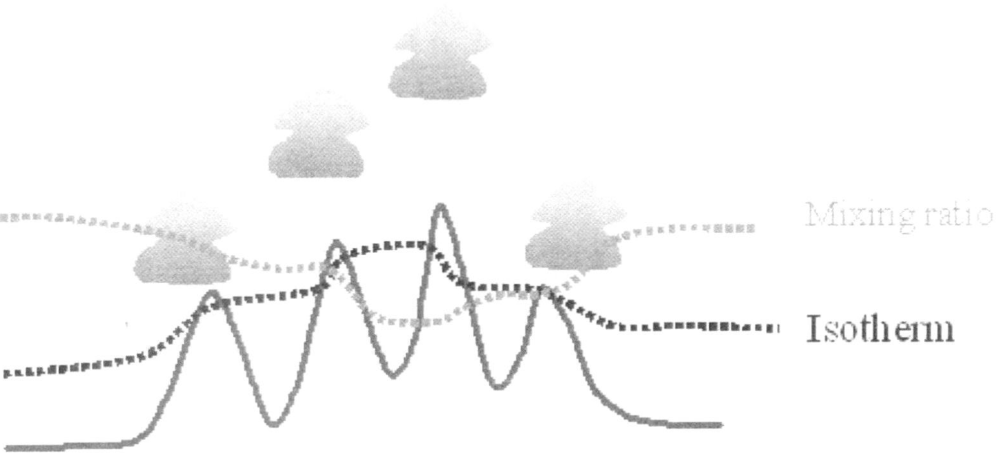

Figure 26. Schematic cross Alpine profile of the temperature, humidity and cumulus condensation level

Although the Alpine Pumping (Lugauer, 2005), visible by the pressure distribution (see Figure 27), suggests a massive Alpine convergence and triggering of convection, this is only concerning low levels. The Anti-valley wind (Whiteman, 2000) at crest level represents a compensating divergent flow above crest level. The important lesson is, that we have to consider the different scales of processes and the vertical structure to be able to understand the role of mountains in triggering convection, there is no simple rule, valid for any case.

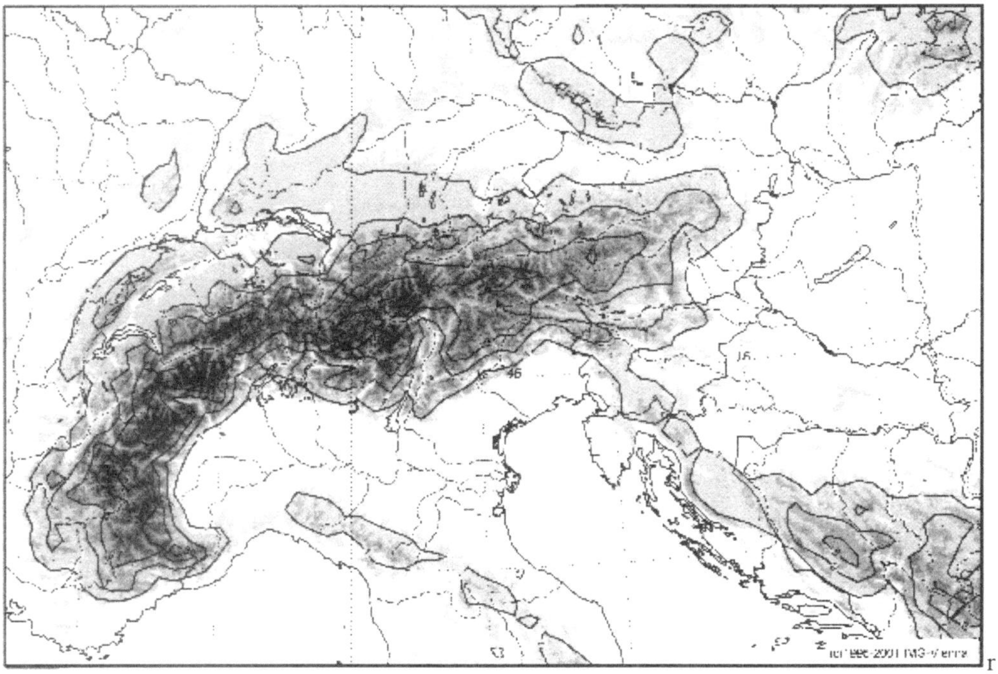

Figure 27. Idealized heat low over the Alps. Isobars are plotted with a 1 hPa interval (after Steinacker et al, 2006)

References

Defant, F (1949) Zur Teorie der Hangwinde, nebst Bemerkungen zur Teorie der Berg- und Talwinde. *Arch Meteor Geophys Bioklim* A1, 421-450.

Egger, J (1990) Thermally forced flows: Theory. In *Atmospheric Processes over Complex Terrain* (W. Blumen, ed.). Meteorological Monographs, AMS Boston, 43-57.

Fliri, F (1975) *Das Klima der Alpen im Raume von Tirol.* Verlag Wagner, Innsbruckm 454pp.

Houze Jr. R A (1993) *Cloud Dynamics* Academic Press, 573pp

Kuetttner, J (1949) Periodiche Luftlawinen. *Meteor. Rdsch.*2, 183

Lugauer, M, P. Winkler (2005) Thermal Circulation in South Bavaria – climatology and synoptic aspects. *Meteor Z.* 14, 15-30

Prandtl, L (1942) *Strömungslehre* Vieweg u. Sohn, Braunschweig, 396-399.

Steinacker, R (1982) Area-height distribution of a valley and its relation to the valley wind. *Cont. Atmos. Phys.* 57, 64-71.
Steinacker, R, M Ratheiser, B Bica, B Chimani, M Dorninger, W. Gepp, C Lotteraner, S Schneider, S Tschannett (2006) A Mesoscale Data Analysis and Downscaling Method over Complex Topography. *Mon Wea Rev*, submitted.
Vergeiner, I (1982) Eine energetiche Teorie der Hangwinde. *Meteor Atmos Phys* N.F. 19. 189-191.
Wagner, A (1932) Der tägliche Luftdruck- und Temperaturgang in der freien Atmosphäre und in Gebirgstälern. *Gerl. Beitr. Geophys.* 37, 315-344.
Whiteman, C D (2000) *Mountain Meteorology.* Oxf. Univ. Pr., 376 pp.
Whiteman, C D, B Pospichal, S Eisenbach, R Steinacker, M Dorninger, P Weihs, E Mursch-Radlgruber, C B Clements (2004a) Inversion breakup in small Rocky Mountain and Alpine Basisns. *J. Appl. Meteor.* 43, 1069-1082.
Whiteman, C D, T. Haiden, B. Pospichal, S. Eisenbach, R. Steinacker (2004b): Minimum temperatures, diurnal temperature ranges and temperature inversions in limestone sinkholes of different size and shape. *J. Appl. Meteor.* 43, 1224-1236.

Topography:
the Global Player in Meteorology

Reinhold Steinacker

Institute of Meteorology and Geophysics, University of Vienna, Austria

Abstract. In this document the basic characteristics of topography on our planet and its impact on the atmosphere is discussed. Fundamental thermodynamic as well as dynamic consequences of mountains with regard to the atmospheric processes are being treated. Relevant scales of phenomena are investigated and their implication on damping or forcing of convection.

1 Geographical Facts of Topography

From the solid earth's surface, roughly 35% are lying above the mean sea level (see figure 1). The mean elevation of this land surface is some 840 m above mean sea level, which means that the mass of a mean air column over land is reduced by 10% against an air column over sea. If we arbitrarily count all areas on land above the mean elevation as mountains, we end up with a percentage of 25% of the land area and 8% of the total earth's surface. This is a considerable amount and explains why mountains play an essential role in the global, regional and local weather processes on earth.

How do mountains influence the atmosphere? Different elevation of a surface means also a different energy budget, radiative as well as turbulent fluxes show significant vertical gradients. This means that elevated surfaces may act as important heat sources or cold sinks, depending mainly on latitude, season and daytime. Also with regard to humidity, mountains may represent significant sources or sinks, leading to increased or reduced precipitation. A mayor influence of mountains on the atmosphere concerns the airflow. The atmospheric flow impinging topography may be deflected vertically or horizontally or eventually be blocked. This modification exerts a strong influence on the humidity budget and hence on the precipitation field. Differential heating and cooling in mountainous terrain produce thermally driven circulations which interact with larger scale circulations. Depending on the sign of energy and humidity budget perturbation, mountains are well known as significant areas of forced or damped convection. If we consider the geographical distribution of the mayor mountain ranges on the earth (Figure 2), we see that on each continent and at each latitude mountains are present. To distinguish between the mountain ranges and their influence on the atmosphere, we have to characterize them: latitude, season, daytime and surface characteristics are the primary factors which explain the radiative budget and hence allow a description of the different behaviour of mountains of comparable size throughout the world. The latitude and season also gives a first estimate on the static stability of the atmosphere. The static stability is the key parameter which allows us in conjunction with the flow speed to

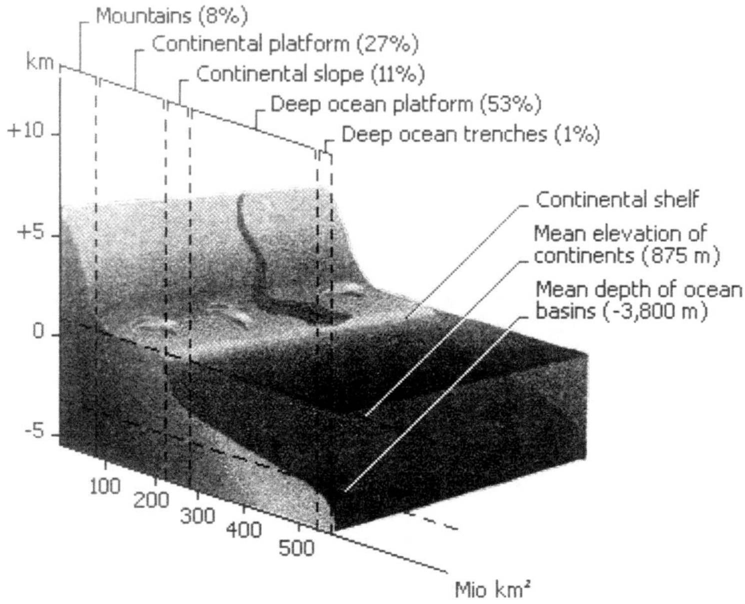

Figure 1. Hypsographic distribution of the earth's surface. The different altitude ranges are given in % of the whole surface (top). The area in millions of km^2 are given on the abscissa, the altitude above and below the mean sea level on the vertical axis (Encarta, 2005)

judge on the possibility of an air parcel to climb a mountain slope or to be deflected or even blocked. The static stability is commonly expressed by the Brunt-Vaisala frequency (see e. g. Reuter et al, 2001), the frequency of a vertically oscillating parcel in a stably stratified atmo-

$$N = \sqrt{\frac{g}{\Theta_0} \frac{\partial \Theta}{\partial z}} \qquad (1.1)$$

sphere. It is important to notice that in practice we determine the vertical gradient of the potential temperature Θ by finite differences which means that an average value over a certain vertical increment Δz is used. If the atmosphere is saturated, the gradient of the potential temperature has to be replaced by the equivalent potential temperature Θ_e. As during a vertical ascent a transition between dry and moist conditions is commonly occurring, the concrete application may become complicated.

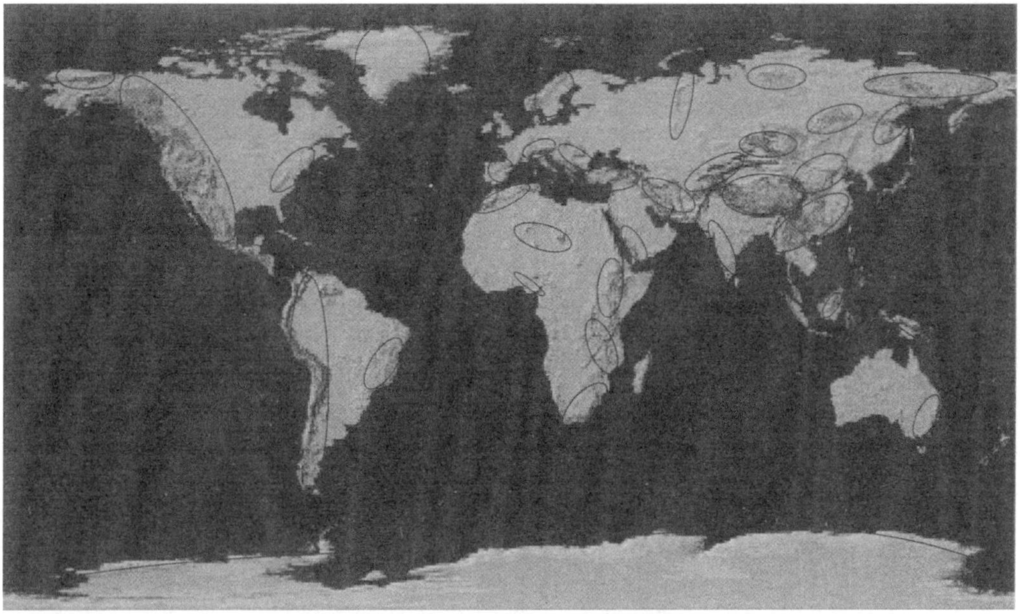

Figure 2. Geographical distribution of the mayor mountain ranges of the world, indicated by ellipses (Peakware, 2005)

A modulation of the influence of mountains on the atmosphere is furthermore exerted by the location with respect to the sea (continentality) and neighbouring mountain ranges. A coastal mountain range with advection of high moisture from the adjacent sea will show a different reaction in producing clouds and precipitation than a continental mountain range, surrounded by dry plains.

2 Flow over versus Flow around an Obstacle

At the same latitude, season, daytime and continentality, the influence of mountains upon the atmosphere is basically determined by the height H. There is no unique rule, which height is to be taken for a specific problem. The effective height is generally different from the maximum height of a mountain massif. Also the area mean of the height of a mountain massif might not be the adequate measure. If one is interested in the dynamical effect of mountains, the mean crest height (relative to the surrounding plains) is the effective height H which is felt by the flow against the obstacle. A measure which is commonly used to determine the question whether the flow is going over or around the obstacle is the Froude number

$$Fr = U / NH \tag{2.1}$$

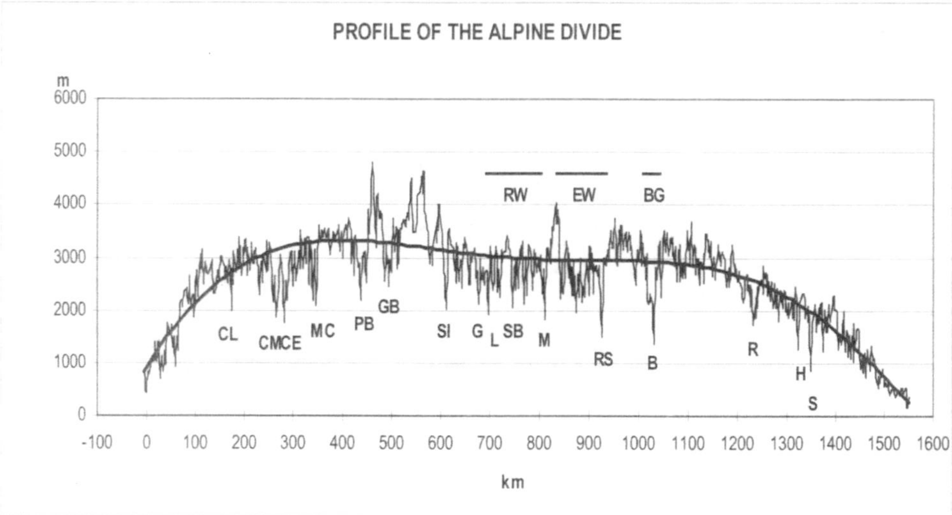

Figure 3. Profile of the Alpine Divide from the Maritime Alps (left) to the Viennese Alps (right). The profile has been produced by using significant point information (height of peaks and passes from high resolution topographic maps). The smooth bold line shows the polynomial fit to the original data. The highest peaks and the lowest passes are more than 1000m above/below the smooth (effective) crest height. The abbreviations mean: CL...Col de Larche, CM...Col de Mont Genevre, CE...Col de l'Echelle, MC...Col du Mont Cenis, PB...Col de Petit St. Bernard, GB...Col de Grande St. Bernard, SI...Passo de Sempione, G...Passo di St. Gottardo, L...Passo di Lucomagno, SB...Passo di San Bernardino, M...Passo di Malora, RS...Reschen Pass, B...Brenner Pass, R...Radstaetter Tauern Pass, H...Hoher Tauern Pass, S...Schober Pass, RW...Rhine valley window, EW...Engadine Window, BG...Brenner Gap, after Steinacker, 2006.

where U is the characteristic speed of the flow against the mountain and N is the Brunt-Vaisala Frequency of equation 1.1. For the European Alps the effective crest height is usually taken close to 3000m, although the highest peak (Mt. Blanc) is more than 4800 m above mean sea level (see figure 3).

The mean crest height in figure 3 shows a slightly higher level for the Western Alps as compared to the Eastern Alps. Considering a standard (dry) static stability $N = 10^{-2}s^{-1}$, the flow against the Alps should be greater than approximately $30ms^{-1}$ to permit a flow over the obstacle. This means that the Alps at a first glance are favourable to block a flow rather than allow the flow going over it.

3 Mountains as Heat Sources and Cold Sinks

If thermodynamic aspects of the influence of mountains on the atmosphere are treated, the effective height of topography is the height above mean sea level on the smallest possible scale. The nonlinear vertical gradients of the single terms of the energy budget equation does not permit a simple areal averaging of topographic heights without making significant errors. If we have a first look to the basic terms of the surface energy budget we come to the following conclusion (see figure 4)

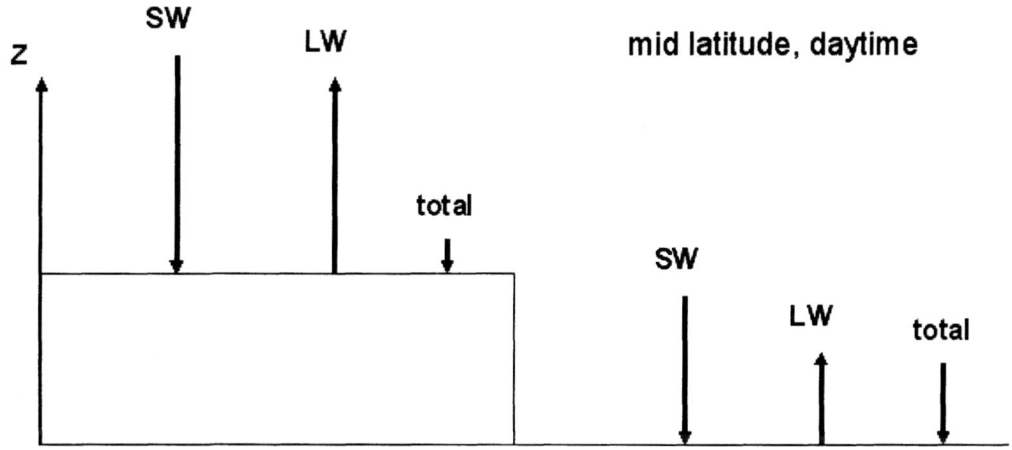

Figure 4. Schematic view of radiation fluxes at different altitudes. SW...net short wave radiation flux, LW...net long wave radiation flux, total...total net radiation flux.

The short wave downward radiation flux is slightly increasing with height if there is no significant change of surface albedo. This is due to the decreasing optical depth of the atmosphere with height and hence the decreasing effect of absorption and scattering. Usually, the upward shortwave radiation flux (reflected radiation) will be more effective for the vertical gradient of the shortwave net radiation. The Albedo may considerably change with height. Difference in surface characteristics, vegetation covered, bare ground or even snow covered ground may produce a decreasing net shortwave radiation flux (pointing downward) despite an increasing downward shortwave radiation flux. The long wave upward radiation flux is typically decreasing with height due to a decreasing surface temperature with height. The downward long wave radiation flux (atmospheric radiation), however, is even more decreasing with height as the water vapour, being one of the basic contributors of atmospheric radiation, is strongly decreasing with height. The net long wave radiation flux (pointing upward) is therefore typically increasing with height. Adding the two – short and long wave net radiation fluxes – gives the total net radiation flux which is pointing downward and shows a general decrease with height. This means that short wave radiative heating is rather decreasing and long wave radiative cooling is rather increasing with height!

Why are then mountains often seen as heat sources? The reason lies in the different vertical temperature gradient of the free atmosphere and the vertical gradient of the equilibrium radiation temperature with height (see figure 5). The standard lapse rate (standard atmosphere value for the troposphere of 0,0065 K/m) which is explained by the moist adiabatic lapse rate leads to a stronger vertical temperature decrease in the free atmosphere as compared to the temperature decrease of the surface radiative equilibrium temperature. This means that on the global average the temperature of an elevated surface is becoming successively warmer with height as compared to the free atmosphere. Of course latitude, season, daytime and ground characteristics may significantly modify this average findings. A further reason for the common description of mountains acting as heat sources is explainable by a different ratio between the sensible and the latent heat fluxes (Bowen ratio, see e. g. Arya, 2001).

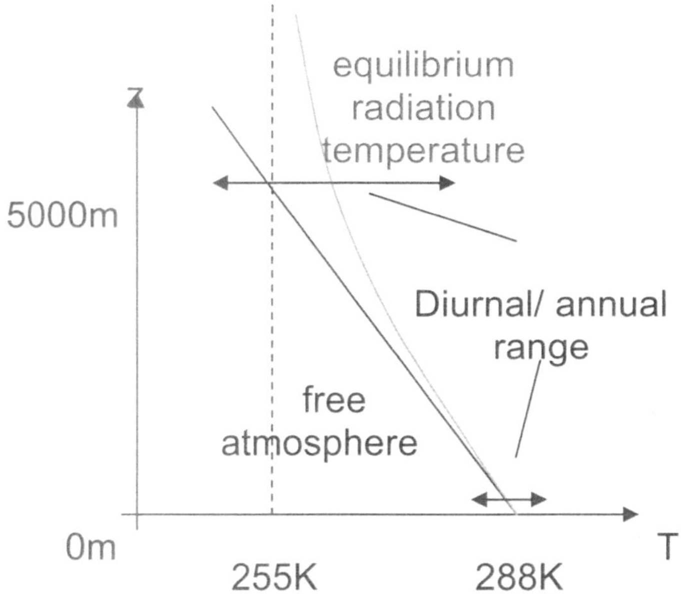

Figure 5. Mean vertical temperature profile of the free atmosphere (black, standard lapse rate of 0,0065K/m) and vertical profile of the radiative equilibrium temperature (grey).

4 Scaling Aspects of Mountains

Besides the effective and mean height of a mountain range other geometric parameters change its influence on the atmosphere. The (half-)width of a mountain range D allows to define the aspect ratio

$$\alpha = H / D \tag{4.1}$$

as a measure of the steepness of the terrain. A steep topography forces the atmosphere in general to climb on the windward side or descent on the lee side steeper than with a smooth sloping barrier. Taking typical values for the European Alps as a whole with $H \sim$ 3km and $D \sim$ 75km the aspect ratio is $\alpha \sim 1/25$, rather smooth. If we, however, take a single Alpine ridge/valley system with $H \sim$ 2km and $D \sim$ 5km we end up with an aspect ratio $\alpha \sim 2/5$, which is fairly steep and explains, why e. g. the channelling effect within Alpine valleys is fairly strong.

Besides the width of a mountain range also its length L is of importance to understand its influence on the atmosphere. Long mountain ranges may prohibit any flow around, i. e. they will block the air on the windward side completely, when no flow over the obstacle is possible. The question if the airflow over mountains may be approximated by a 2D approach may be answered by a further ratio: the ellipticity of a mountain range

$$\varepsilon = L/2D \tag{4.2}$$

By definition $L \geq 2D$. If $L \gg 2D$, mountains are very long – as compared to their width – and can be treated as a two dimensional barrier. For the European Alps with $L \sim$ 750km and $2D \sim$ 150km the ellipticity $\varepsilon \sim 5$. This means that the European Alps cannot really be described by a 2D approach. It has to be furthermore noted, that the Alps consist of two segments, the mayor West-East oriented and the minor North-South oriented part, which prohibits a precise indication of ellipticity. In connection with the ellipticity the orientation of a mountain range with respect to the dominant flow is important to be considered. If a mountain range in the Westerlies is oriented West to East, there is hardly a pronounced windward and lee side, e. g. documented by a wet and dry side. As most part of the Alps is West-East oriented, this is also true for this mountain range. Both sides show approximately the same amount of annual precipitation.

Of high importance - besides the aforementioned characteristic parameters to describe mountains - the 3D structure plays an eminent role. If a mountain crest is not straight but curved, which is commonly the case, the mountain drag is strongly dependent on the flow direction (see figure 6).

If a flow is directed from the concave side of a mountain range (left side of figure 6) the flow, in case of low Froude number, has to change its direction by more than 90 degrees to flow around the obstacle. In the opposite case (right side of figure 6) the directional change is below 90 degrees. Hence the windward pressure perturbation, which is necessary for the direction change has to be larger in the first case.

A further important factor which explains a different influence of mountains on the atmosphere is the subscale structure. In figure 3 the along crest profile of the Alps is shown. The variability of this profile may have significant effects on the cross mountain flow. We do not only have the possibility of a flow around versus flow over but we may also distinguish between flow "through" the mountains. This means that on a small scale the flow across the deep cut mountain passes – below the mean height of the crest - may lead to significant local phenomena. Gap flows permit a certain portion of the blocked windward air to flow to the lee side with all dynamical and thermodynamical consequences. As some Alpine Passes are cut into the Alpine crest by more than 1000m, gap flows or shallow foehn are very common phenomena there.

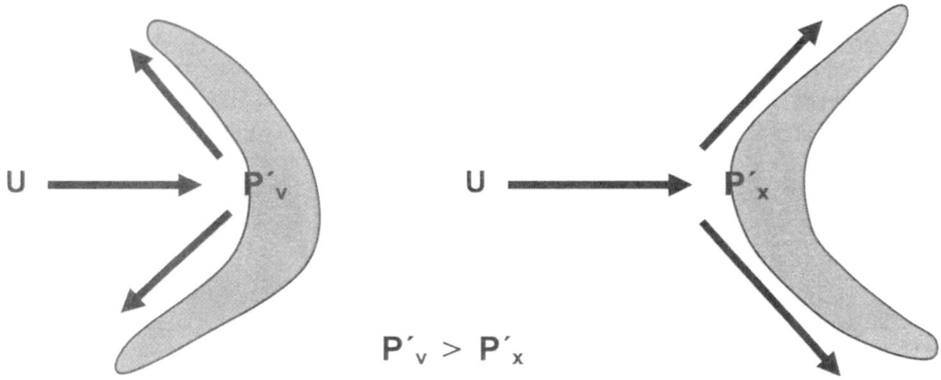

Figure 6. Schematic picture to show the different pressure drag of a concave (left) and a convex (right) mountain range. Further explanation see text.

Often, when treating mountains and their dynamical and thermodynamical role, a simple analytic profile is chosen, the witch of Agnesi (see e. g. Durran, 1990)

$$h = H\frac{D^2}{D^2 + x^2} \tag{4.3}$$

where h is the height in profile across a mountain range with half width D and crest height H as a function of distance x of the crest.

This allows to derive some very useful analytic solutions for the airflow over topography. One should however keep in mind that real topographies do not always fit to such simple assumptions. Some prototypical cross profiles of mountains are shown in figure 7.
It can easily be comprehended that the different profiles of figure 7 mean a different dynamical and thermodynamical response of the atmospheric forcing due to mountains.

So far only for the Alpine characteristic scales have been considered. If we look at the variety of mountains of the world, we can see that a whole range of scales is present. Taking the Orlanski, 1975 classification, we find mountains on the Large scale (20.000 - 2.000km), e. g. L of the Andes, the Rocky Mountains, or $L \sim 2D$ of the Antarctic plateau. On the Meso-α scale (2.000 – 200km) several mountain ranges are matching like L of the European Alps, of the Appalachians, the New Zealand Alps or the width (2D) of the Greenland plateau or the Rocky Mountains. On the Meso-β scale (200 – 20km) many mountain ranges are matching their width ($2D$), like the European Alps, the Appalachians, the New Zealand Alps and many more. The Meso-γ scale (20 – 2 km) is the typical width scale ($2D$) for single mountain ranges within mountain massifs around the world, also volcanic cones are typically lying in this range. The smallest scale, the Micro scale (2km – 200m) is typical for single hills and single mountain peaks around the world.

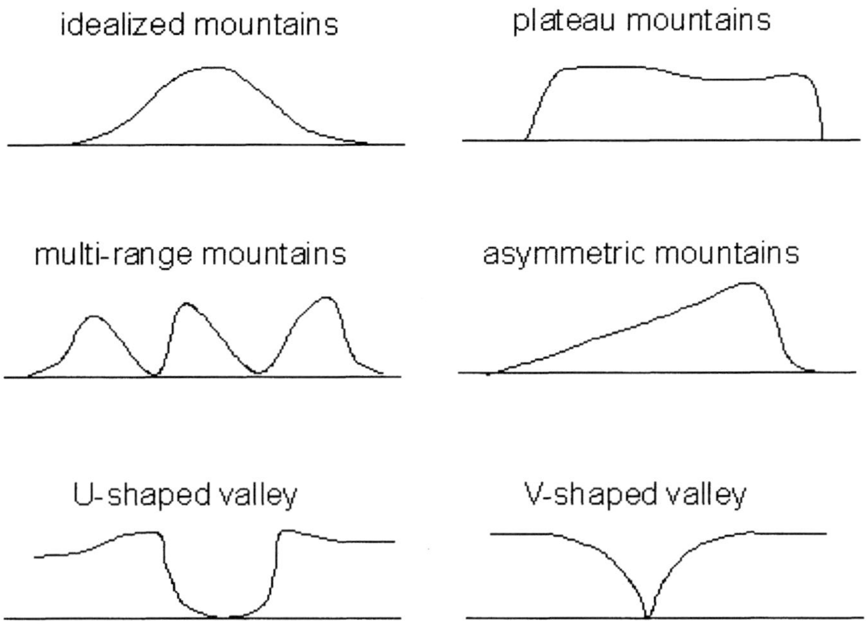

Figure 7. Prototypical cross mountain/topography profiles.

With a very few exceptions, mountains, including their substructure, are commonly ranging over several scales, which means that scale interaction becomes an important issue, when treating the influence of mountains on the atmosphere. A view at the topography of the European Alps (see Figure 8) shows the different scales of topography clearly. The Alps as a

Scale	Shallow convection	Single deep convection	Mesoscale convective system	Alpine massiv	Alpine valley/ridge
L	1km	10km	100km	750km	150km
2*D*	1km	10km	100km	150km	10km
H	1km	10km	10km	3km	2km
U	1m/s	10m/s	10m/s	10m/s	10m/s
W	1m/s	10m/s	10m/s	U*H/2D=0,2m/s	U*H/L=0,1m/s
T	0,1h	1h	10h	D/U=2h	L/U=4h

Table 1. Typical scales of convective phenomena and Alpine orographic phenomena. L is the length scale, D the (half-)width scale, H the height scale, U the characteristic horizontal flow speed, W the characteristic vertical wind speed and T the time of the life cycle of convection and the duration of the flow over the topography and within a valley respectively.

whole have a length L of approximately 750km (meso–α), and a width $2D$ of roughly 150km (meso–β). The mayor Alpine valleys and ridges have a length L on the average of say 150km (meso–β) and a width $2D$ of say 10km (meso–γ). A comparison of the scales of convection with the scale of the Alps is given in table 1.

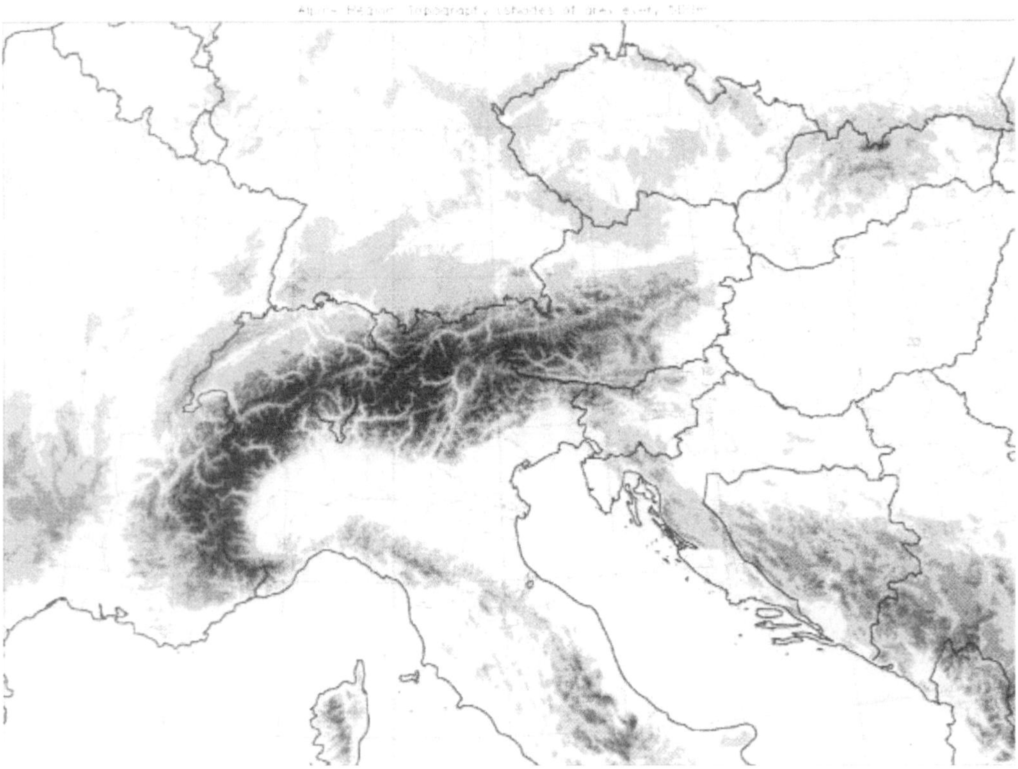

Figure 8: Topography of the European Alps and its surroundings, based on a 1 km resolution. The shading starts at 500m above mean sea level and becomes darker every further height increment of 500m.

Comparing the scales in tab. 1 it is evident that convection is more related to cross Alpine flows and valley flows than flows along the mountains. A severe implication of this finding means that e. g. the resolution of a prediction model and its topography becomes a very important issue with regard to a correct treatment of topographically induced convection.

References

Arya, S. P. (2001) *Introduction to Micrometeorology*. Academic Press, 450 pp.

Barry, R. G. (1992). *Mountain Weather and Climate.* Routledge (UK). 432 pp.

Durran, D. (1990). Mountain Waves and Downslope Winds. In: *Atmospheric Processes over Complex Terrain* (W. Blumen, ed.), Meteorological Monographs, AMS, Boston, 59-81

Encarta (2005). Internet: http://uk.encarta.msn.com/media.

Orlanski, I (1975). A rational subdivision of scales for atmoapheric processes. *Bull Amer Meteor Soc* 56, 527-530.

Peakware (2005), Internet http://www.peakware.com/encyclopedia/zoom.htm

Reuter, H., M. Hantel, R. Steinacker (2001). Meteorologie. In Bergmann Schäfer, Lehrbuch der Experimentalphysik, *Bd. 7. Erde und Planeten.* (W. Raith. Ed.). Walter de Gruyter, Berlin New York. 131-310.

Steinacker, R (2006), Alpiner Föhn – eine neue Strophe eines alten Liedes. *Promet: Gebirgsmeteorologie.* Deutscher Wetterdienst, Offenbach, in print.

Environmental Conditions Associated with Convective Phenomena: Proximity Soundings

Harold E. Brooks

NOAA/National Severe Storms Laboratory, Norman, OK, USA

Abstract. An important tool in understanding the relationship between environments and observed severe thunderstorm events are vertical profiles of environmental conditions collected in the vicinity of the storms. These relationships can help in the future forecasting of weather. In this paper, the use and cautions associated with these so-called proximity soundings are discussed.

1 Background

One of the fundamental tenets underlying efforts to forecast severe weather is the assumption that the environmental conditions in an area provide useful information to help determine what weather occurs. This underlies the premise behind the notion of ingredients-based forecasting. If we know what ingredients would produce a particular weather event, then observing those ingredients in a location is a strong indication that we can expect the weather event in question. Because of that, studies of what kinds of conditions are observed with specific classes of weather events have been carried out in order to learn about the relationships. Note that the studies don't exactly match the forecast problem. If, for instance, a set of conditions exists that will produce a particular weather event, but that those conditions occur together very rarely, then studies of the conditions associated with observed events are unlikely to observe them. As a result, they may not impact the forecast problem. Nevertheless, the studies are likely to provide information that is useful for forecasting.

One of the primary classes of approaches to the problem is the study of so-called "proximity soundings", in which radiosonde observations taken in the vicinity of an event are studied. This has been used extensively in the studies of severe thunderstorms and tornadoes. The first study was that of Beebe (1958). A long history of studies has followed. For details, see Brooks et al. (1994), Rasmussen and Blanchard (1998), and Rasmussen (2003) and references therein. Much of what is now used for forecasting severe thunderstorms and tornadoes has been derived from these studies.

Brooks et al. (1994) contain a discussion of the challenges in defining "proximity." In general, one would like to have environmental conditions that represent the atmosphere that produced the storm. In practice, this presents some difficulties. In most studies, proximity is defined by putting limits on the distance between the storm and sounding in both time and space with, often, some constraints that attempt to eliminate soundings that are not representative of the storm environment. The time and space constraints explicitly recognize that the atmosphere is not homogeneous. The variation in time and space is large enough that the event and the sounding have to close to each other, but small enough that the sounding actually does represent the environment of the storm. Clearly, these two conditions make it challenging to construct reasonable definitions. Of course, any values of the constraints are arbitrary. The difficult issue of how to eliminate soundings that are within the time and space constraints, but still not representative of the environment, has no easy resolution. For small samples, it is possible to consider every sounding individually for things like convective contamination or for being on the wrong side of a large-scale boundary.

The issue of sample size for the any dataset of soundings is at the heart of the decision-making process on how to design the study. Larger time and space constraints will, obviously, increase the number of soundings in the dataset, but increase the likelihood of the sounding being unrepresentative. For large samples, any procedure to eliminate unrepresentative soundings within the time and space constraints will have to be automated. For forecast applicability, this has benefits, as well as problems. Obviously, soundings may make it in to the dataset that are not representative and hurt the quality of the relationship developed between the weather and sounding. On the other hand, it is frequently difficult to be confident that a sounding is unrepresentative of the environment and, in practice, the forecasting problem is exactly one of knowing that the environment tells the forecaster what they need to know to forecast what will occur. Inclusion of soundings that are not obviously unrepresentative in gross ways makes the proximity study much more like the real forecasting problem. As such, it may increase the utility of the research for forecasters.

Rasmussen and Blanchard (1998) [hereafter RB98] attempted to solve some of the problem with defining proximity by using a flow-based definition. Soundings were included if the sounding represented the air mass going towards the storm. Thus, closer soundings might be thrown out and more distant soundings used in some cases depending on the wind direction. This should eliminate some of the problems with being on the wrong side of boundaries. It is not clear if the effect is sufficiently large to make a large difference in the results compared to more traditional approaches.

Recently, Thompson (1998), Markowski et al. (2003), and Thompson et al. (2003) have expanded the use of proximity soundings to numerical model soundings. These approaches may mimic the forecast process in many ways, since the ability to estimate what the environmental conditions will be should help the forecast. Thus, forecasters could learn what the relationship between forecast atmospheres and observed weather looks like. Brooks et al. (2003) used global reanalysis soundings to create proximity datasets. This has the advantage of producing consistent soundings in time and space that are not convectively contaminated. It has the disadvantage of adding the question of the relationship between the actual atmosphere and the reanalysis into the mix. It may be more useful for addressing climate-scale questions and, in fact, the main thrust of the paper was to produce global estimates of the frequency of severe weather environments.

Caution must be used in constructing proximity studies, particularly if the goal is to identify the conditions that are associated with a particular event from another event that the soundings are chosen appropriately. If, for instance, one is interested in identifying strong tornado environments, simply collecting tornado soundings and comparing the conditions to randomly selected soundings, it is likely that the dominant differences that will be seen will be between those variables that identify thunderstorm environments (CAPE will be the dominant parameter). If the comparison is between strong tornadic and non-tornadic thunderstorm soundings, the dominant differences will be between supercell and non-supercell environments (a combination of CAPE and deep shear, as will be seen later). Comparing strong tornadic and non-tornadic supercell soundings will provide information on the differences in tornadic and non-tornadic soundings, given that the soundings are associated with supercells. Since it appears that discriminating supercell from non-supercell environments is relatively easy, the comparison of the tornadic and non-tornadic soundings will focus on different parameters.

2 Lessons Learned From Proximity Studies

By far, the most extensive proximity studies of severe thunderstorms have been carried out in the US. Some of the basic results will be discussed here. RB98 carried out a study of an entire year's worth of soundings with non-zero CAPE in the US, giving a larger sample size than previously used. They identified three classes of events: those with strong (F2 or stronger) tornadoes (tornadic), those without strong tornadoes, but with large hail (supercells), and those with no large hail or strong tornadoes (ordinary). Their results are representative of many of the large datasets that have been created. Strong thunderstorms (supercells or tornadic) could be discriminated from ordinary thunderstorms using the CAPE of the environment with the CAPE defined using a parcel mixed over the lowest 1 km of the atmosphere (Fig.1). Only 25% of the ordinary soundings had CAPE greater than approximately 1100 J kg^{-1}, while half of the other soundings did. There is little, if any discriminatory capability for CAPE between the tornadic and supercell environments. There is a slight tendency for the highest part of the distribution for tornadic CAPEs to be skewed farther towards high values. While there are small differences in the 75th percentile values, the 90th percentile for tornadic soundings is approximately 25% higher than for supercells. It's possible that such differences are physically significant, but since the values are in the tails of the distributions, they're not very common. As such, even if the difference is "real", it may not have practical significance in forecasting.

RB98 carried out a similar analysis for the wind shear, using the magnitude of the difference between the mean boundary layer wind and the 6 km wind. Again, the shear can be used to discriminate between the ordinary thunderstorms and the stronger storms (Fig. 2). The 50th percentile value for ordinary storms is 10.8 m s^{-1}, which is less than the 25th percentile for the other two categories. The median supercell and tornadic shear values are skewed towards the upper portion of the 25th-75th percentile range, while the median ordinary shear is more nearly centrally located. Thus, high shear is a strong predictor for supercell storms, given the presence of CAPE. As with the CAPE, deep tropospheric shear is not particularly effective at discriminating between tornadic and supercell soundings. The median and 75th values of the supercell soundings are slightly higher than the tornadic, but the extreme values for the tornadic soundings are more extreme. As a result, the potential for use of shear by itself as a forecast parameter to discriminate between the stronger storms is limited.

Combining the shear and CAPE in a two-dimensional display improves the discrimination between supercells and non-supercells. Craven and Brooks (2005) produced a dataset similar to RB98, but used three years worth of data. Their definition for the intermediate category was slightly different, including hail with 5 cm diameter or wind gusts of at least 120 km h^{-1}. Strong thunderstorms are much more likely when both CAPE and shear are strong (Fig. 3). It might be possible to put lower bounds on both CAPE and shear, below which strong thunderstorms are very unlikely (say, 7 m s^{-1} and 300 J kg^{-1}). There appears to be a slight tendency for the strong tornadic soundings to be associated with stronger shear than the non-tornadic storms. Very few tornadic soundings are found with shear below 20 m s^{-1}. In any event, there is no obvious discrimination possible between the tornadic and non-tornadic severe soundings using these deep parameters. The deep parameters appear to do a reasonably good job of discriminating between

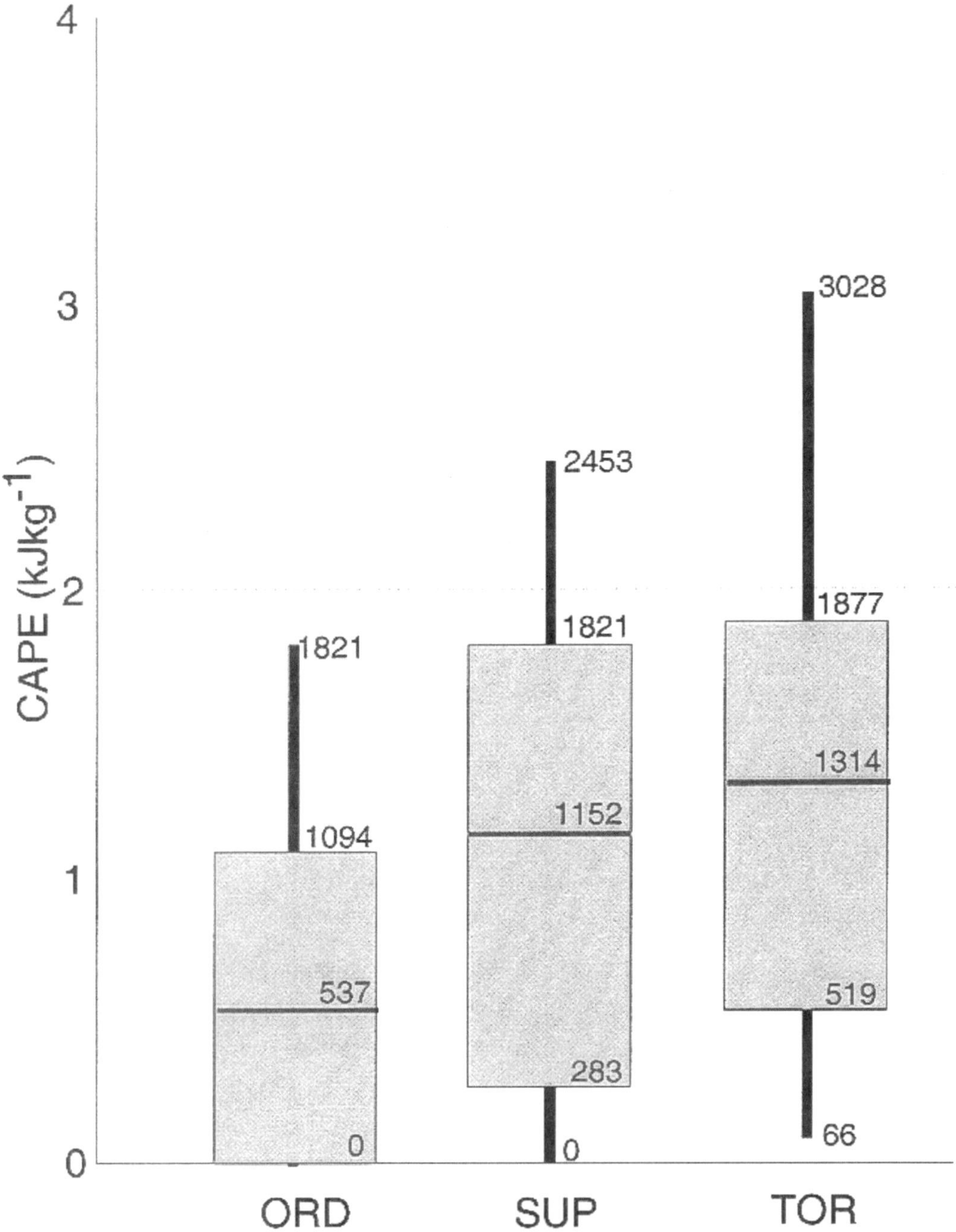

Figure 1: Box and whiskers graph of CAPE for soundings associated with supercells with significant tornadoes (TOR; right), supercells without significant tornadoes (SUP; middle), and nonsupercell thunderstorms (ORD; left). Gray boxes denote 25th to 75th percentiles, with heavy horizontal bar at the median value. Thin vertical lines (whiskers) extend to the 10th and 90th percentiles. From Rasmussen and Blanchard (1998).

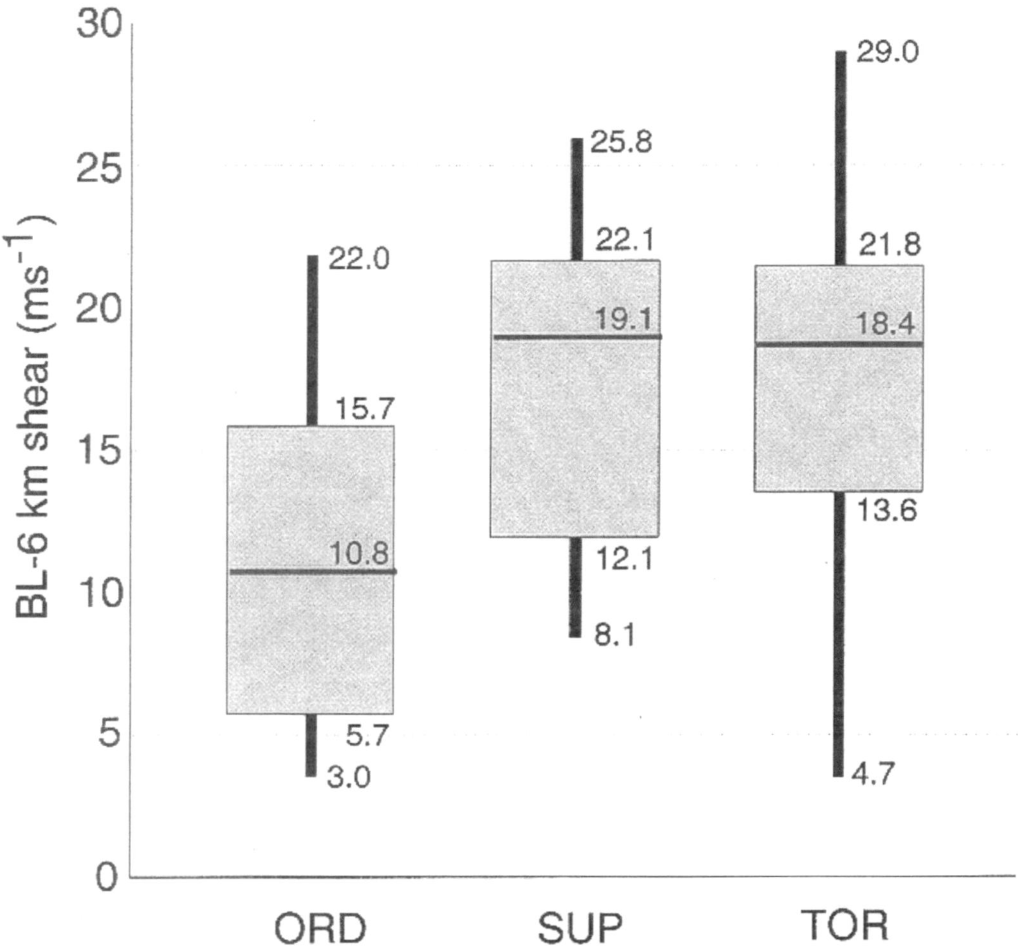

Figure 2: Same as Figure 1, except for magnitude of difference between mean boundary layer and 6 km above ground level winds. Also from Rasmussen and Blanchard (1998).

the strong severe (or strong tornadic) environments and the ordinary environments. Thus, they can be thought of as a first cut for forecasting the presence of supercells, assuming that some initiation mechanism for thunderstorms can be identified. Going beyond that to identify tornadic environments requires looking at different environments and larger datasets. The Craven and Brooks (2005) dataset contains fewer than 600 supercell soundings. In a two-dimensional display, the 600 soundings are relatively small. To amplify the dataset, Brooks et al. (2002) went back and found additional soundings associated with significant severe thunderstorms back to 1973, yielding a dataset of 2460 soundings.

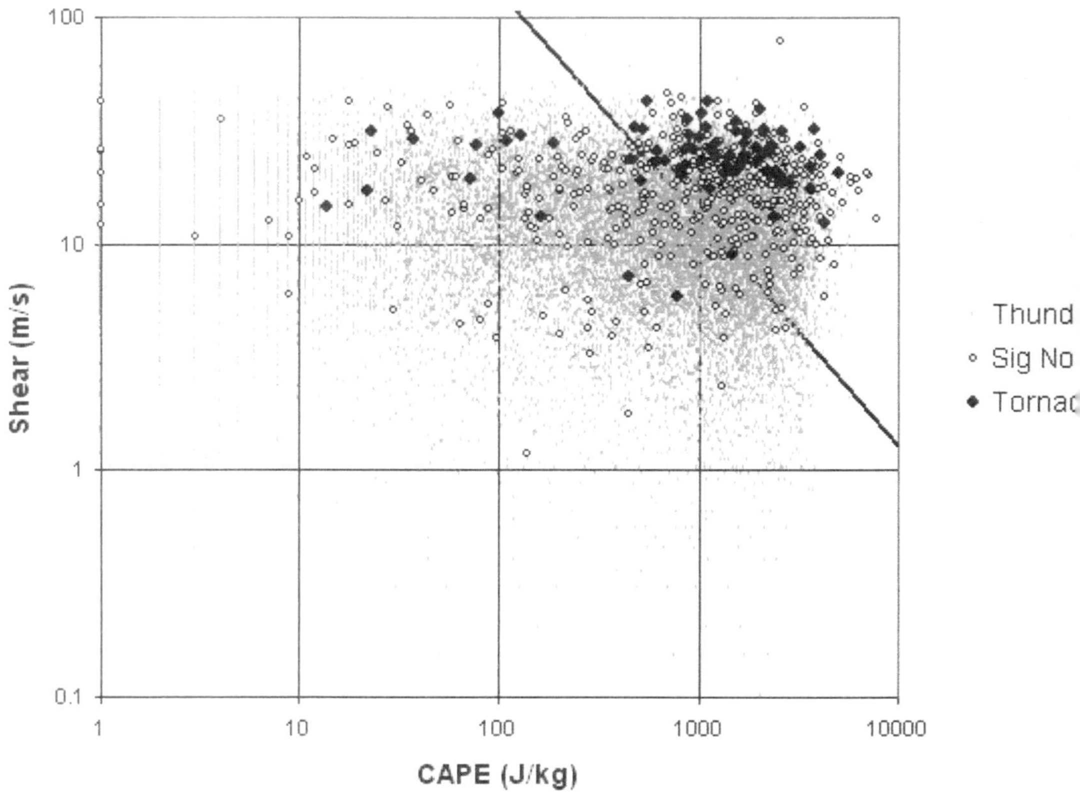

Figure 3: CAPE and magnitude of 0-6 km wind difference ("Shear") for 0000 UTC soundings for US from 1997-1999. Black diamonds are associated with F2 or stronger tornadoes, open circles with 5 cm or larger hail or 120 km h^{-1} wind gusts without F2 or stronger tornado and small gray dots are associated with thunderstorms that don't fit in other categories ("ordinary" thunderstorms). Black line is line of best discrimination between ordinary thunderstorms and stronger thunderstorms, based on linear discriminant analysis.

The results of the analysis of tornadic vs. non-tornadic severe storms are illustrated in Fig. 4 (Brooks and Craven, 2002, also presented in Brooks, 2005). Tornadic soundings are much more likely with high values of low-level shear and low lifted condensation levels (LCL). Further analysis (not shown) indicates that this is particularly true of soundings taken east of 95° W in the US (roughly the longitude of Oklahoma City). West of there, tornadoes are still more likely with high shear and low LCL heights, but the difference between the probabilities are smaller. There is some suggestion that the extreme CAPE values found in the Plains of the US are related to the difference in the regions. If the highest CAPE values are excluded from the analysis, the two regions begin to look more alike, although there is still some difference.

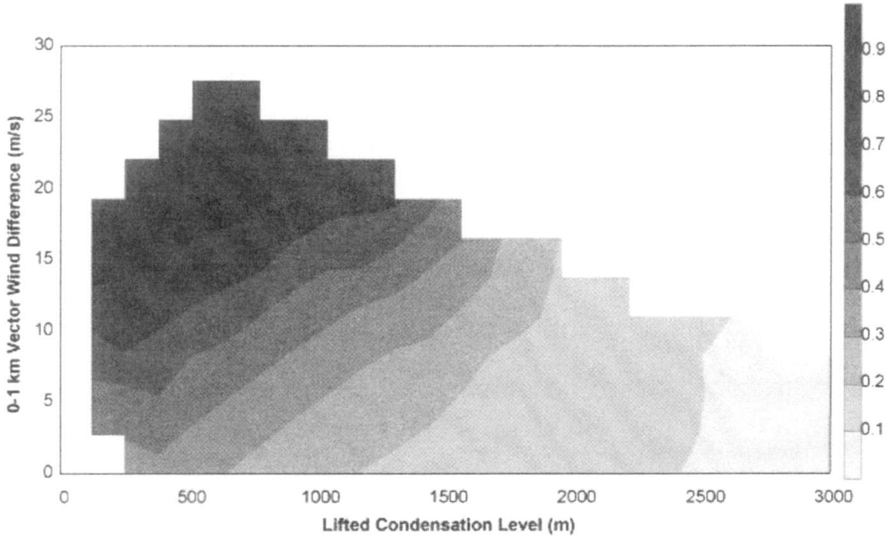

Figure 4: Probability of significant tornado given that a storm produces significant severe weather. Lifted Condensation Level (LCL) calculated using a parcel mixed over lowest 100 hPa of atmosphere. Dataset consists of 2460 soundings, including 740 tornadic soundings, taken within 180 km and 3 hours of severe weather event from 1973-1999 in the US. Observed conditions were gridded on this domain (grid spacing 125 m in LCL, 2.5 m s^{-1} in wind difference) and assigned to location on grid if within an ellipse with radius 275 m in LCL, 7.5 m s^{-1} wind difference. Probability calculated by dividing number of tornadic soundings at grid location by total soundings. Value only calculated if at least 30 soundings assigned to grid point. Figure from Brooks and Craven (2002) and also presented in Brooks (2005).

In order to attempt to look globally, Lee (2002) constructed soundings from the NCAR/NCEP reanalysis. In the US, those soundings discriminated between ordinary thunderstorms and stronger thunderstorms in the same way as the observed soundings, lending credibility to their use for other purposes. The discrimination between significant tornadic and significant non-tornadic soundings was not as good in the reanalysis as in the observations. Brooks et al. (2003) used the soundings in order to estimate global severe thunderstorm frequency and noted, in particular, that the estimated tornado maximum in the US was shifted to the east. They speculated this was due to the discrimination technique picking up on tornado outbreak situations because of their strong signal, biasing the results. At the 2004 European Conference on Severe Storms, Romualdo Romero presented a list of approximately 80 significant tornadoes from Europe. Thirty-eight of the soundings had proximity soundings already created from the reanalysis. A comparison of the distribution of CAPE and deep shear for the Romero dataset and US soundings (Fig. 5) shows that the European soundings tend to be shifted towards lower CAPE and shear. If, however, only those soundings occurring in the eastern part of the US (east of 95° W) on days when only one sounding in the country is in proximity to a significant tornado, the shift is much less. It appears that the Plains of the US and large tornado outbreaks in the US occur in parts of the CAPE-Shear phase space that European soundings never (or very rarely) visit. Thus, the US proximity data may provide insight into European conditions, but care must be taken in applying the data.

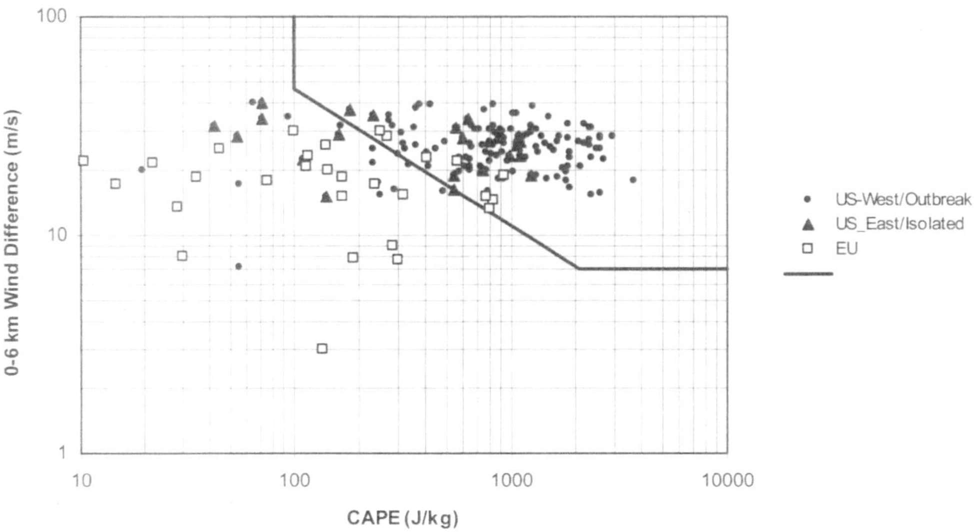

Figure 5: CAPE and deep shear parameters associated with proximity soundings near F2 or stronger tornadoes taken from NCAR/NCEP reanalysis. Black dots are for US tornadoes west of 95° W or with soundings taken on days with more than one proximity sounding collected for significant tornado, black triangles are for US tornadoes east of 95° W on days when only one proximity sounding was collected, and open squares are for European tornadoes. Black line is same as in Figure 3 except with cut-offs at 7 m s^{-1} shear and 100 J kg^{-1} CAPE added.

As discussed above, discrimination between tornadic and non-tornadic significant severe storms appears to be done best using a combination of the LCL height and low-level shear. Here, the comparison between the European and US data is a little different (Fig. 6). In the eastern US isolated tornado events, high LCL storms are not seen in the dataset. It appears that weak low-level shear is not as much of a problem for the European soundings. Whether this is due to a different combination of the appropriate ingredients (assuming that we even have a complete list), or with difficulties with the reanalysis depiction of the environment is unknown at this time. In both the eastern US and Europe, the relatively high LCL (>~1 km), high shear environments aren't sampled. In the western US and outbreak sample, they are relatively common. From looking at other aspects of the data set, they appear almost entirely to be associated with very high CAPE environments. It is possible that this is a reflection of the greater potential for stretching of vorticity associated with the very high CAPEs, although that is speculation at this time. It does appear, though, that a set of ingredients occurs in the plains and during outbreaks that aren't observed in other locations very frequently, if at all.

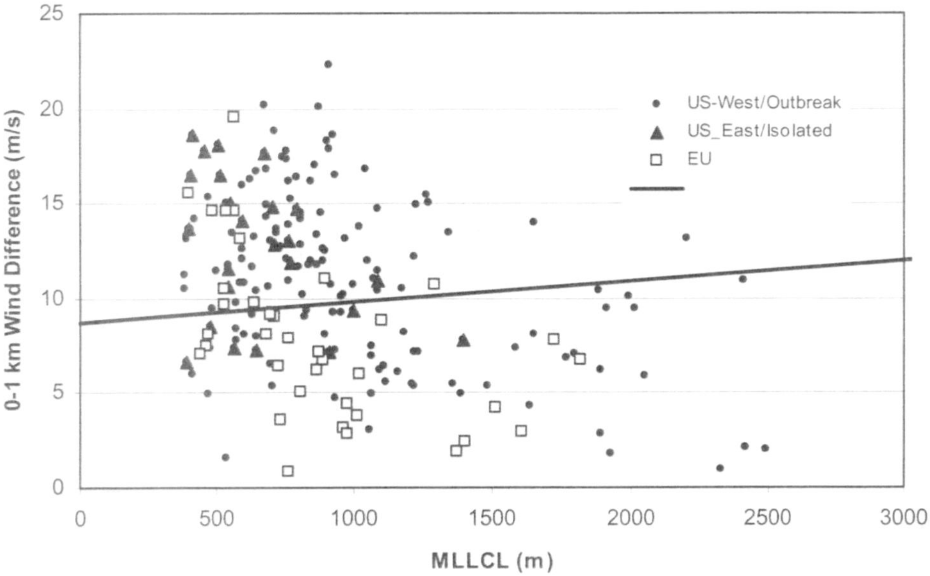

Figure 6: Same as Figure 5, except for lifted condensation level of sounding using a mixed layer 100 hPa deep (MLLCL) and for the difference in the surface and 1 km above ground level winds.

3 Summary

Proximity data analysis can be a valuable tool in understanding environmental conditions for severe thunderstorms. They can help in understanding what kinds of atmospheric conditions are typically observed during severe thunderstorm events. On another level, they help aid forecasters in knowing when unusual conditions are occurring. In many areas of human endeavor, it is difficult for people to estimate the frequency of rare events. In weather forecasting, it is extremely important for forecasters to recognize unusual conditions when they come together. A rare set of atmospheric conditions at a location is almost certainly going to be associated with a rare weather outcome. Proximity sounding studies can provide information for forecasters on this point. Creation of large data sets, frequently requiring looking over a large area and time, as well as local studies of the frequency of occurrence of conditions (e.g., Manzato, 2005) can be a great aid in both regards.

References

Beebe, R. G. (1958). Tornado proximity soundings. *Bull. Amer. Meteor. Soc.* 39: 195-201.

Brooks, H. E. (2005). Ingredients-based forecasting. This volume.

Brooks, H. E., and J. P. Craven (2002). A database of proximity soundings for significant severe thunderstorms, 1957-1993. *Preprints*, 21st Conference on Severe Local Storms, San Antonio, Texas, American Meteorological Society, 639-642.

Brooks, H. E., C. A. Doswell III, and J. Cooper (1994). On the environments of tornadic and non-tornadic mesocyclones. *Wea. Forecasting* 10: 606-618.

Brooks, H. E., J. W. Lee, and J. P. Craven (2003). The spatial distribution of severe thunderstorm and tornado environments from global reanalysis data. *Atmos. Res.* 67-68: 73-94.

Craven, J. P., and H. E. Brooks (2005). Baseline climatology of sounding derived parameters associated with deep, moist convection. *Nat. Wea. Digest,* in press.

Craven, J. P., R. E. Jewell, and H. E. Brooks (2002). Comparison between observed convective cloud-base heights and lifting condensation level for two different lifted parcels. *Wea. Forecasting* 17: 885-890.

Lee, J. W. (2002). Tornado proximity soundings from the NCEP/NCAR reanalysis data. M. S. Thesis, University of Oklahoma, 61 pp.

Manzato, A. (2005). The use of sounding derived indices for a neural network short-term thunderstorm forecast. *Wea. Forecasting,* in review.

Markowski, P., C. Hannon, J. Frame, E. Lancaster, A. Pietrycha, R. Edwards, and R. L. Thompson (2003). Characteristics of vertical wind profiles near supercells obtained from the Rapid Update Cycle. *Wea. Forecasting* 18: 1262–1272.

Rasmussen, E. N., and D. O. Blanchard (1998). A baseline climatology of sounding-derived supercell and tornado forecast parameters. *Wea. Forecasting* 13: 1148–1164.

Rasmussen, E. N. (2003). Refined supercell and tornado forecast parameters. *Wea. Forecasting* 18: 530–535.

Thompson, R. L. (1998). Eta model storm-relative winds associated with tornadic and non-tornadic supercells. *Wea. Forecasting* 13: 125–137.

Thompson, R. L., R. Edwards, J. A. Hart, K. L. Elmore, and P. Markowski (2003). Close proximity soundings within supercell environments obtained from the Rapid Update Cycle. *Wea. Forecasting* 18: 1243–1261.

Development and Use of Climatologies of Convective Weather

Harold E. Brooks

NOAA/National Severe Storms Laboratory, Norman, OK, USA

Abstract. Estimates of the occurrence ("climatologies") of convective phenomena in time, space, and intensity can be useful in a variety of contexts. They provide background for forecasters, and the risk management and meteorological research communities. In part, because of the different needs of those user groups, caution must be applied when developing and using climatologies, especially if the intended application is outside of the original intent of the developers.

1 Goals of Climatological Development

Severe thunderstorms are a threat to life and property all over the world, with the possible exception of Antarctica. Climatological studies of their occurrence serve a variety of purposes for a range of people. In this paper, I'll consider the value of these studies and how they can be developed and utilized.

In the broadest sense, climatologies can be thought of as the description of what kind of weather occurs where and when. Note that the "where" and "when" can be related to the environmental conditions in which the storms form as well as the spatial and temporal description. The "what" can include descriptions of the intensity or other factors describing the event.

For forecasters[1], climatological information can provide important background for making forecasts. Knowing what conditions and what events are common or uncommon can provide a starting point for making a forecast. The climatological information can help in calibrating the forecasts. If a forecaster always overestimates (or underestimates) the amount of rain, for instance, climatological information can help to remove the bias from the forecasts. If the climatological information involves relating environmental conditions to observed weather, it is also possible that the information will improve the forecast of events that are very rare at a particular location but for which information has been gathered in other locations. On a related issue, based on the American experience, the beginning of efforts to forecast severe thunderstorms corresponded with an increase in collection of reports of severe thunderstorms. In effect, people don't forecast what they don't think occurs and they don't make an effort to find out how often things happen if they don't believe the events happen. As a result, there is a positive feedback between forecasting for a kind of weather event and the quality of the observational database that allows us to know how often the event occurs.

For researchers, the information can help put bounds on the extent of their knowledge of a subject. Being able to establish relations between events in time and space and the state of the atmosphere represents a high bar to be cleared. From a statistical standpoint, being able to

[1] In this context, "forecasters" can refer to human beings or numerical models or any other system or combination of those aspects that generates a forecast.

construct models that produce output that have the same statistical properties as the record of observations would allow one to estimate the true hazards from severe weather events. At that point, stochastic modelling can provide insight into the nature of the phenomenon, including important information on the interannual variability and probability of rare events.

The risk management community, including emergency managers and insurance concerns are also interested in evaluation of the hazards. They have long-term planning needs in terms of allocating resources and setting rates in response to threats.

Finally, estimation of changes that might be associated with possible climate change scenarios can only be made in the context of an understanding of the current situation. Appropriate climatological studies can put bounds on what kinds of changes can be detected, and how much data would be required to make such detections.

2 Data Limitations

Ideally, one would like to have a long record of consistently collected and recorded data in order to construct climatologies. From the statistical standpoint, such a record allows for the calculation of various moments of the distribution. Consistent collection means that meteorological trends can be detected. A common example of such a record is the temperature observation at a particular site. Except for changes in the area around the site (urbanization, change of exposure, etc.) or changes in policy, the data record can be recorded consistently for a long period of time at a particular location.

Severe thunderstorm reports differ from such a record in almost every possible way. With the exception of isolated locations with networks of hailpads to collect hail data, the US has the most aggressive system for data collection on severe weather. Thus, in many ways, it serves as an upper bound on the quality of reporting databases. Comments on its structure should provide insight into the nature of the problems.

Reports initially enter the US system from a variety of sources. Historically, starting 1916, the forerunner of the National Weather Service (NWS) produced monthly summaries of tornadoes from their local offices, picked up mainly from newspaper reports. (As an aside, later researchers extended those records back using newspaper reports.) Starting in 1953, when the NWS began forecasting tornadoes, they started to collect reports at the local offices in near-real time, adding hail and convective wind gusts in 1955. In 1975, the NWS adopted the Fujita damage scale for tornadoes (Fujita, 1981). The NWS then retrospectively rated tornadoes in the database back to 1950. There is evidence to suggest that the retrospective rating process overrated tornado damage (Verbout et al., 2005). Other changes have occurred to the database over the years, including a requirement to rate all tornadoes on the Fujita scale, even if there is no information on which to base a rating, and a requirement added in the mid-1990s to require offices to specify either a value for the speed of the wind gusts or amount of dollar damage for wind damage, eliminating reports of nonspecific wind damage with no wind speed estimate. Unfortunately, this has dramatically increased the number of reports of winds at 50 kts, the lowest possible severe wind speed.

Presently, reports can come from the general public, public safety officials, trained severe weather spotters, or from observations. They go through the local NWS office and are vetted by the Warning Coordination Meteorologist (WCM) who is responsible for putting the local data together for a monthly report. It is important to note that the primary purpose of collecting the reports is for the verification of warnings issued by the office, not for producing high-quality datasets of observations. Since warnings are typically valid for a county for at least 30 minutes, multiple reports within those time and space concerns may not get included in the final dataset.

It is not surprising that there is large variability in time and space in the reports. Individual offices, even now, may have different standards for collecting the reports, resulting in a patchwork quilt that is unlikely representative of meteorological differences (Doswell et al., 2005). In addition, over the years, emphasis on reporting, changes in population and road structure, new technology (especially radar), and changes in public awareness have all resulted in differences in the quality of the reports over time (Brooks et al., 2003; Doswell et al., 2005, hereafter B03 and D05, respectively). An example of the effects of these changes and the problems associated with interpretation of the record is seen in the annual number of tornadoes reported in the US (Fig. 1). There are a number of features in the record that could be of interest meteorologically. Overall, there has been a huge increase in the number of reports over time. A big question is how much, if any, is meteorological. Looking in detail, an increase in 1953-1954 is likely the result of reports being collected in real time. (The question of whether the big change occurs in 1953 or 1954 is open to debate. Other evidence, including consideration of the number strong tornadoes, supports the idea that 1953 was a better year for tornadoes than 1954, so it is hard to determine exactly what's going on.) Occasional years (e.g., 1957) stand out above the long-term trend of increasing reports. The 1970s and 1980s look relatively stationary. In recent years, 1998 and 1999 look like big years, 2002 is low, and 2004 is the biggest year on record. In large part, the 2004 record is a result of several large outbreaks associated with landfalling hurricanes on top of an otherwise normal year. Verbout et al. (2005) fit a linear regression to the 1954-2003 data and adjusted the data to that regression, a technique that makes the late 1980s look like a period of low tornado events and the early 1970s look like a period of high tornado events. That would be consistent with anecdotal reports from long-time tornado chasers (C. Doswell, personal communication) and consistent with the pattern of number of reports of strong tornadoes. However, this illustrates the fundamental problem. Significant meteorological changes (say a 25% increase) could be masked by the changes in reporting practices.

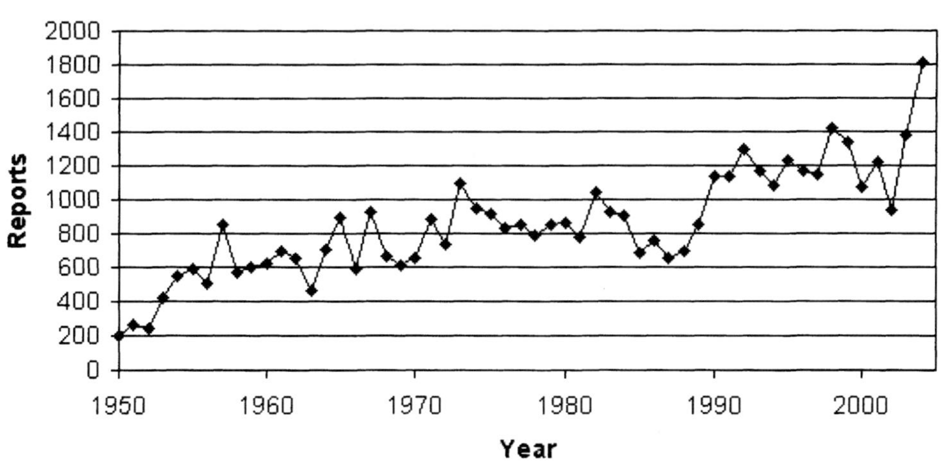

Figure 1: Annual reports of tornadoes in the US tornado database, 1950-2004.

This simple illustration shows the primary challenges associated with interpreting severe weather databases. What aspects of the databases are consistent over time and space? How can we take advantage of those consistencies? What are the limits on what can be said? In the following sections, I'll show examples of some of the things that have been done to try to deal with the problems. Fundamentally, however, the limit is the availability of high-quality data.

3 Redefining the Problem

One way to approach the problem of small datasets is to redefine what the climatology is for. As an example, B03 and D05 considered the climatology of severe weather "days," rather than the severe weather event. This reduced the effects of the "inflation" in reports seen in Figure 1. The problem then became the distribution of how many days that severe weather occurred at a location, rather than how many severe weather events occurred. Obviously, if many severe weather events occurred someplace on a single day, the impression would be different in the two approaches.

Another technique is to include information from other locations. Again, B03 and D05 gridded reports on an approximately 80x80 km grid, including information from a relatively small area, but then smoothed the grid using a Gaussian smoother. The effect was to use information from a larger area, retaining the primary influence from the central location. The choice of the scale of the spatial smoother was dictated by the goals of the exercise, in this case to match the scale of Storm Prediction Center forecasts. It was hoped that some of the spatial inhomogeneities in the reporting process would be washed out by the smoother.

In the spirit of taking information from other areas, looking at the distribution of reports by intensity can also be useful. Many kinds of weather events follow simple statistical distributions. For severe thunderstorms, which by definition are rare events, we're frequently interested in the tales of distributions. Extreme value distributions look like exponential distributions in the tails, so that on log-linear plots, the distribution of events by intensity would lie on a straight line. Many physical situations (size of volcanic eruptions vs. number, size of forest fires vs. number) follow power-law distributions or truncated power-law distributions. For a power-law distribution, the number of events exceeding a threshold (a cumulative distribution) versus intensity, usually measured in terms of energy, falls on a straight line on a log-log plot. Upper-truncated power laws fall off rapidly after following a power law over relatively low ranges of energy. Other distributions also provide simple plots, such as the apparent Weibull distribution for tornadoes by Fujita scale (Dotzek et al., 2003). The distribution falls off more rapidly in some locations, such as those dominated by non-supercell tornadoes, but are consistent in other locations (Brooks and Doswell 2001.)

These kinds of plots can identify consistent features in the distribution and can, in fact, identify bad data. Individual events can't be identified as bad, but the cumulative effect can be seen. Brooks and Stensrud (2000) found that the distribution for high precipitation amounts followed a log-linear distribution over a broad range (Fig.2). It fails at the high end, indicative of problems in the data set.

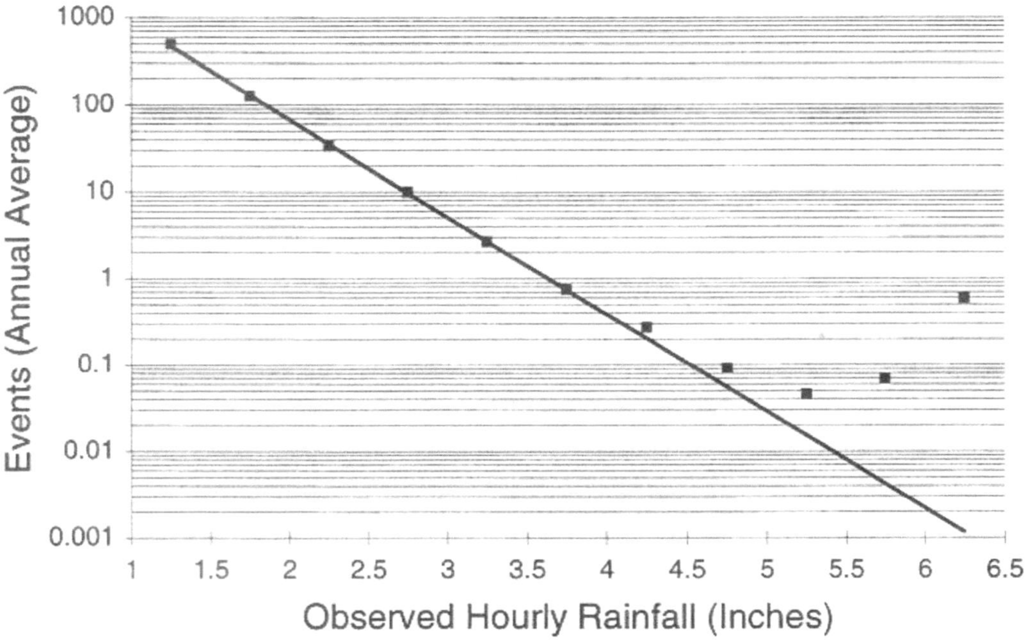

Figure 2: Number of reports of hourly precipitation per year from the US Hourly Precipitation Dataset from Brooks and Stensrud (2000). Point in 6-6.5 inch range includes all values above 6 inches.

Wind gust reports tend to follow power-law distributions (Fig. 3). In the case of the US data, one of the things that is apparent is the underreporting of low end severe wind gusts in the earlier period of record. Above the lowest values, the slopes of the two distributions are similar, implying that the values could have been taken from the same distribution. A similar result, although on a log-linear plot, can be seen in the hail reports by size (Fig. 4). The hail reports illustrate another issue. In the US, there are certain sizes of hail that are reported more often than others. One of those is related to the sizes of sports balls (golf balls, baseballs, softballs) to which hail is compared. Very little hail is reported to be just small or just larger than golf balls, but much is reported right at golf balls. To take care of that problem, the hail can be binned into approximate sizes, in this case, near integer values of inches. It is likely true that there is some underlying distribution that, if reports were collected with greater precision, is the "true" distribution. The distribution that we observe is a result of the filtering process of making the reports. Preliminary analysis (not shown) indicates that the distribution is less steep (falls off slower, so that large sizes are more likely) in locations where CAPE is higher. This makes some physical sense, but verifying it requires larger datasets, since restricting the analysis to high CAPE or low CAPE regions necessarily limits the size. Differences in the distributions for wind gusts don't show up with a simple pattern in the wind data. More analysis is needed in order to understand this.

Datasets that don't follow simple distributions should be treated with some caution. Small samples can lead to unusual distributions, but certain kinds of problems can be detected. In comparing the distribution of tornadoes by intensity for France and for the pre-1950 US database,

Brooks and Doswell (2001) noted that the distributions were similar and, using the fact that US reports had dramatically more weak tornadoes once the real-time reporting began, surmised the French reports would increase as well if real-time reporting began.

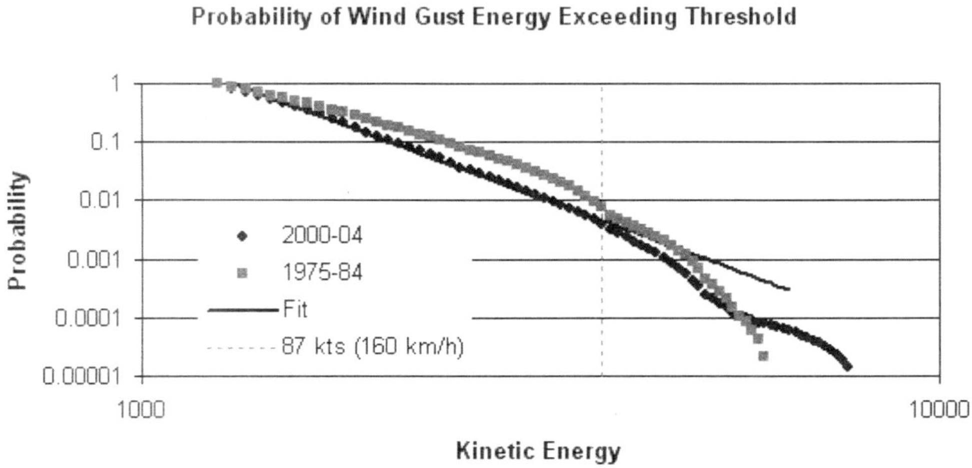

Figure 3: Distribution of reports of wind gusts versus kinetic energy associated with reports in the US. Points are probability that kinetic energy exceeds value of energy. Black points are for reports from 2000-2002, blue are for 1975-1984. Straight lines indicate power law distribution for wind gusts from approximately 90-120 km h^{-1}.

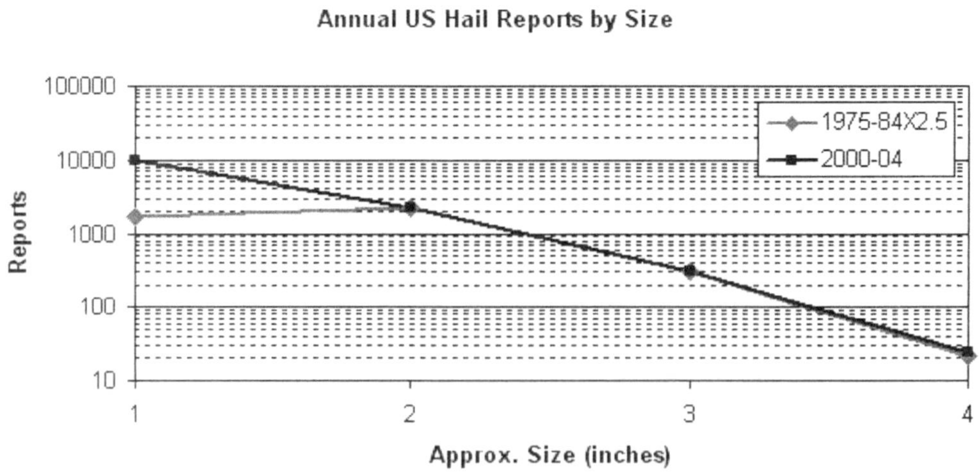

Figure 4: Annual number of hail size reports for US. Black line is for period 2000-2003, blue is for reports from 1975-1984, multiplied by 2.5. Hail sizes have been binned to nearest inch.

4 Environmental Conditions

Another approach to looking at climatology is to consider the distribution of environmental conditions associated with convection. Relationships between the environment and events can be constructed using proximity sounding procedures (Brooks, 2005). Brooks et al. (2003b) used relationships based on the NCAR/NCEP reanalysis to make estimates of the world-wide distribution of severe thunderstorms and tornadoes. Because of our interest in the relationship between the environment and weather, determining the distribution of background environmental conditions is of value. Using the NCAR/NCEP reanalysis data, annually-averaged cycles of important convective parameters have been computed for the eastern US and for Europe. The results, including animations of the annual cycle are available at <URL: http://www.nssl.noaa.gov/users/brooks/public_html/climloop.html>. Even though there are problems with the US reporting database, it is certainly better and more consistent than the European database. As a result, differences in the distribution of environments in the two regions can be important in understanding expected severe weather.

One of the most important differences is the lack of steep lapse rates in Europe. In fact, the lapse rates show neither the high or low extremes in the US. A sample plot from one time during the year for the mid-level lapse rates is given in Fig. 5. In the southern plains of the US (in the Oklahoma vicinity), high lapse rate air coming off of the southern Rocky Mountains overlays high values of lowest 100 hPa mixing ratio (not shown). As a result, we would expect very high values of CAPE to occur often. The importance of the Rocky Mountains and Gulf of Mexico in defining the convective environments of the US is clear. Southerly flow off the Gulf provides a relatively continuous source of low-level moisture. The Rockies provide a source of high lapse rate air. As that air mass moves eastward, the probability of convection is higher than for lower lapse rate air. As a result, convection in the central US consumes the high lapse rate air and the atmosphere in the northeastern US has lower lapse rates than any place in Europe. By the end of summer (not shown) the US lapse rates in the northeast get even lower, so that the gradient in lapse rate is very large across the nation.

5 Summary

Climatological studies provide background information for forecasts, helping forecasters make a first guess for what will occur in the absence of other information. More importantly, probably, they help us understand the relationship between large-scale parameters and weather events. That, in turn, helps in applying understanding developed in locations where reporting databases are reasonably good to places where the reporting databases are of a lower quality.

References

Brooks, H. E. (2005). Environmental conditions associated with convective phenomena: Proximity soundings. This volume.

Brooks, H. E., C. A. Doswell III, and M. P. Kay (2003a). Climatological estimates of local daily tornado probability. *Wea. Forecasting* 18: 626-640.

Brooks, H. E., and C. A. Doswell III (2001). Some aspects of the international climatology of tornadoes by damage classification. *Atmos. Res.* 56: 191-201.

Brooks, H. E., J. W. Lee, and J. P. Craven (2003b). The spatial distribution of severe thunderstorm and tornado environments from global reanalysis data. *Atmos. Res.* 67-68: 73-94.

Brooks, H. E., and D. J. Stensrud (2000). Climatology of heavy rain events in United States from hourly precipitation observations. *Mon. Wea. Rev.* 128: 1194-1201.

Brooks, H. E., and S. J. Weiss (1999). Severe local storms. *Proceedings of the WMO/UNESCO Sub-forum on Science and Technology in Support of Natural Disaster Reduction* WMO-No. 914, 12-31.

Doswell, C. A. III, H. E. Brooks, and M. P. Kay, (2005). Climatological estimates of daily local nontornadic severe thunderstorm probability for the United States. *Wea. Forecasting* 20:in press.

Dotzek, N., J. Grieser, and H. E. Brooks (2003). Statistical modeling of tornado intensity distributions, *Atmos. Res.* 67-68: 163-187.

Trapp, R. J., S. A. Tessendorf, E. S. Godfrey, and H. E. Brooks (2005). Tornadoes from squall lines and bow echoes. Part I: Climatological distribution. *Wea. Forecasting* 20: 23-35.

Verbout, S. M., H. E. Brooks, L. M. Leslie, and D. M. Schultz (2005). Evolution of the U.S. tornado database: 1954-2003. *Wea. Forecsting*, submitted.

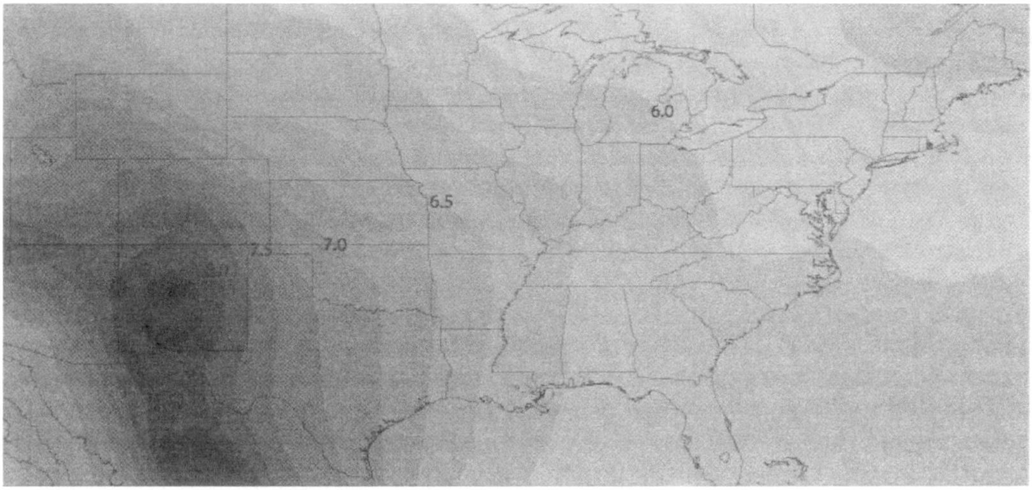

700-500 hPa Lapse Rate on 21 May

700-500 hPa Lapse Rate on 9 July

Figure 5: 700-500 hPa Lapse rate (shaded every 0.25 K km^{-1}, with labels every other contour) for eastern US on 21 May (top) and Europe on 9 July (bottom). Color scheme is same for both figures.

Ingredients-Based Forecasting

Harold E. Brooks

NOAA/National Severe Storms Laboratory, Norman, OK, USA

Abstract. Forecasting the weather can be thought of as a problem in extracting a small signal from a noisy background field. Much information is available to the forecaster, but, frequently, only a small amount of that information is of importance for solving the forecast problem(s) of the day. As a result, an approach to forecasting must maximize the efficiency of the process. An effective way, particularly for hazardous weather, is to identify the ingredients required to produce a particular weather event and then to focus on the processes that can affect the presence of those ingredients. This allows the forecaster to narrow the range of aspects of the observations and model guidance that are considered during the forecast shift and, it is hoped, identify crucial developments as they occur.

1 Background

Forecasting the weather is a challenging exercise. It is a low signal-to-noise ratio problem and frequently, it seems that there is both too much information and too little information to make an accurate forecast with a great deal of detail. Severe thunderstorms can provide particular challenges, given their often rapid development, relatively small spatial extent, and the penalties that may be associated with missing events. Historically, forecasting has often relied on "rules of thumb," empirically-derived relationships between observations or some weather event and another weather event. Rules of thumb can range in sophistication from simple weather folklore or sayings (e.g., "Red sky at night, sailor's delight; red sky in morning, sailor take warning.") to elaborate discussions that have the appearance of scientific backing (e.g., "if the wind is out of the west at station X and the heights fall at station Y, expect rain at Z".) In the absence of an understanding of physical cause-effect relationships, empirical rules can be helpful in forecasting, but they are limited to situations that have occurred often enough to have rules derived for them. Pattern recognition is, in effect, an organized set of rules of thumb that depends on the applicability of the analogue pattern that is being compared. A more general approach is the ingredients-based approach (Doswell et al., 1996), which attempts to use the environmental conditions necessary for a weather event as the prime inputs into forecasting.

At this point, it is useful to distinguish between ingredients and parameters. An ingredient is an intrinsic property of the atmosphere. A parameter is some measure of the magnitude of that ingredient. Thus, absolute humidity is a parameter describing moisture, which is an ingredient for convection. The usefulness of parameters in forecasting is related to whether they describe the basic ingredients for a weather event of interest and how strongly they relate to the ingredients. In all cases, it behooves forecasts to know how the parameter they are using is derived and how it relates to the ingredients.

2 Ingredients for Severe Convection

A scientific approach to forecasting involves identifying the ingredients that are necessary for any weather event and then focusing on the processes that can change the distribution of those ingredients and, thus, change the probability of the event occurring. Obviously, the success of this approach depends on the knowledge of the forecaster about the ingredients and the ability to identify the processes. Unfortunately, for many severe weather events, our understanding of the ingredients is imperfect. Thus, forecasters can be put in the difficult position of having to issue forecasts when the state of knowledge is less than perfect. In the remainder of this section, I'll attempt to describe the basic ingredients of convective weather.

In one sense, the primary concern is with the forecasting of deep, moist convection. From that point, the questions about the nature of weather associated with the convection can be addressed by considering the specific ingredients beyond the basics that are believed to be required for the weather event.

Doswell et al. (1996) discuss the three basic ingredients needed for deep, moist convection. They are instability, moisture, and lift. Instability is the presence of an environmental lapse rate that is conditionally unstable. A saturated parcel will rise along a moist adiabat. As such, we need the environmental lapse rate to be steeper than the moist adiabat. This can be measured by the lapse rate, typically in the mid-troposphere, although the appropriate level may depend on the situation. Moisture is related to the amount of water vapor in the near-surface layer. There must be enough moisture so that a rising parcel of air has a level of free convection (LFC), at which point it will be warmer than its environment and rise without being forced. Moisture can be expressed by a number of quantities, such as the specific humidity or dew point. It is important as the primary source of energy for convection, from the potential for energy release associated with the change of phase of water vapor to liquid. Describing the moisture is also complicated by the question of what the appropriate layer over which to mix the moisture. A surface parcel will typically produce higher values than a layer mixed over some deeper layer. Craven et al. (2002) showed that using a parcel mixed over the lowest 100 hPa makes more accurate predictions of lifted condensation levels than a surface parcel, with the surface parcel biased too low. Unfortunately, that does not mean that a 100 hPa mixing layer is appropriate for all applications, but it does imply that the choice of parcel is important and forecasters need to be aware of what parcel is used. The final ingredient for deep, moist convection is a mechanism to lift the parcel to the level that the instability can be realized. This lifting mechanism can take a variety of forms, including orographic lifting and boundaries in the atmosphere, such as fronts, dry lines, and outflow boundaries from previous convection. The primary point of the lifting mechanism is that it must be concentrated on the mesoscale, rather than synoptic scale (Doswell, 1987). Synoptic-scale lift can destabilize the atmosphere, but sufficiently strong lift to initiate convection must be on a shorter horizontal scale.

Note that the order does not imply anything about the relative importance of each one. All are necessary and it is possible to have two of the three and not get convection. As an example, the atmosphere could have significant moisture and instability, but nothing to lift parcels of air to their LFC. Such a situation is common in summer in the central part of the United States where the atmosphere may be "capped", with an inversion at the top of the boundary layer that is strong enough to require massive lifting. Also, in passing, it is important to note that the instability and moisture can be combined and be described by the buoyancy. The vertical integral of the buoyancy is the convective available potential energy (CAPE). See Markowski (2005a) for more discussion of buoyancy and CAPE.

Identification of the ingredients for thunderstorms is important, but additional ingredients are needed for severe convection. Of particular importance to the severe thunderstorm forecast problem are supercell thunderstorms, characterized by persistent, rotating updrafts (Markowski, 2005b). Supercell thunderstorms are prolific producers of hail, strong winds, tornadoes, and heavy precipitation (Smith et al. 2001). A large number of studies (e.g., Rasmussen and Blanchard 1998) showed that CAPE and strong wind shear over most of the troposphere are important for supercell production. The shear acts to organize and prolong the storm. Brooks et al. (2003) showed that the most severe thunderstorm events (presumably associated with supercells) are much more probable in high CAPE-shear environments than in low CAPE-shear environments. The rest of the section is devoted to the ingredients for specific severe thunderstorms. Another summary has been given by Brooks and Weiss (1999).

2.1 Tornadoes

The most significant tornadoes are associated with supercell thunderstorms. Our ability to forecast the intensity is limited, at best, but observational studies (e.g., Rasmussen and Blanchard 1998) have shown that there is better discrimination between severe storms that produce tornadoes that are rated at least F2 on the Fujita scale and those that don't, than if the threshold is any tornado versus nontornadic storms. This may be a result of a special set of physical processes being required to produce the stronger tornadoes, requiring a combination of factors to be in place in the atmosphere. Our understanding of the details of the tornadogenesis process is limited, although many advances have been made in the last thirty years (Markowski 2005c). As a result, our ability to describe the ingredients in such a way that we have unambiguous ways to quantify them for forecasting is limited. Nevertheless, analysis of a large dataset of soundings taken within 180 km and 3 hours of significant severe thunderstorms (producing at least a tornado with F2 damage, hail of 5 cm diameter, or wind gusts of 120 km h^{-1}) collected in the US from 1973-1999 provides some guidance (Brooks and Craven 2002). A combination of two parameters, the magnitude of the difference between the winds at the surface and at 1 km above the ground and the height of the lifted condensation level provides a reasonable discrimination. High shear and low LCL heights are clearly favored.

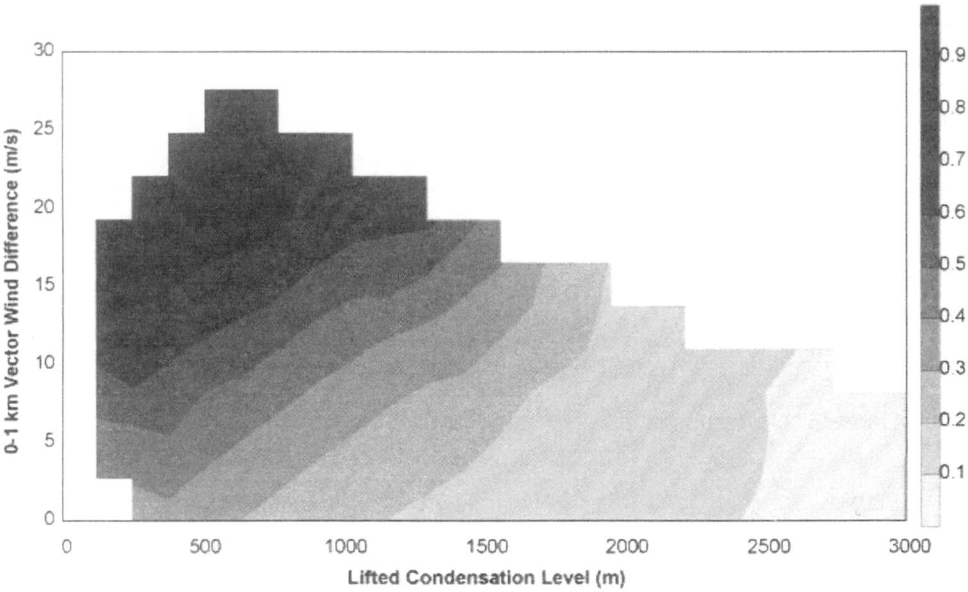

Figure 1: Probability of significant tornado given that a storm produces significant severe weather. Lifted Condensation Level (LCL) calculated using a parcel mixed over lowest 100 hPa of atmosphere. Dataset consists of 2460 soundings, including 740 tornadic soundings, taken within 180 km and 3 hours of severe weather event from 1973-1999 in the US. Observed conditions were gridded on this domain (grid spacing 125 m in LCL, 2.5 m s^{-1} in wind difference) and assigned to location on grid if within an ellipse with radius 275 m in LCL, 7.5 m s^{-1} wind difference. Probability calculated by dividing number of tornadic soundings at grid location by total soundings. Value only calculated if at least 30 soundings assigned to grid point.

2.2 Hail

The problem with hail forecasting is somewhat different than for tornado forecasting. We understand much of the basic physics associated with hail growth. The problem is that we don't observe the relevant variables on a routine basis. Updraft strength, storm-scale wind structures, microphysical behavior, and melting effects as the stone falls all play a role in the creation and size of hail reaching the ground. Of those, updraft strength (via the relationship between CAPE and updraft strength discussed in Markowski (2005a)) is the only one that is relatively well-estimated from routine environmental observations. It appears that a small amount of CAPE perhaps as little as 100 J kg^{-1}, is required to produce some hail. Underlying the observational problem is that the identification of severe hail depends on where and when the hail occurs. Some agricultural interests are susceptible to damage from any hail, while urban residents may not notice damage until the hail gets to be a few cm in diameter. This is further complicated by the fact that, at different times during the growing season, vulnerability may change. A little hail falling on a field of wheat that is ready to harvest will cause significantly more damage than much more hail falling on the same field prior to plant germination. A final question relates to susceptibility to a large amount of small hail versus a few gigantic stones. Hail can drift to depths of half a meter, a la snow, and these drifts can cause major problems, such as blocking drainage, allowing accompanying rainfall to pool and causing significant transportation problems in urban areas. A parked automobile might not receive much damage from the hail, if it is all on the order of a centimeter in diameter, but would likely suffer significant damage if struck by a single 10 cm diameter stone. Urban flooding might be less in such a case.

Efforts to forecast hail size have traditionally been limited to a small number of parameters and have suffered from sample size problems. Forecast experiments at the Storm Prediction Center in the US (Brooks 2005) have shown little or no skill in simple techniques. It appears that the most accurate scheme to date is that of Brimelow et al. (2002), which takes environmental profiles from observations or numerical weather prediction models and runs them through a simple cloud model. Apparently, the model is sophisticated enough to capture the non-microphysical processes well enough to provide some information on size. This is done, however, at the cost of significantly higher computational effort than the simple techniques. Unfortunately, the simple techniques have not shown sufficient skill to make them useful.

2.3 Winds

Strong convective wind gusts can be generated in different types of synoptic environments. Fundamentally, we are looking for the development of strong downdrafts, which will produce strong horizontal winds when they reach the surface. This implies that parcel paths on a thermodynamic diagram have to be descending on a line that falls to the left of the temperature profile. A simple way is for a parcel that is cooled below the temperature aloft could descend dry adiabatically if the environmental lapse rate is nearly dry adiabatic as well. Another way is for a parcel to remain saturated (and hence fall along a moist adiabat) as it falls in an unsaturated environment. Obviously, this requires the time scale of the drying of the parcel to be longer than it takes the parcel to fall.

It is not uncommon for thunderstorms that form in weak wind shear to produce strong wind gusts. In most cases, attention has to be paid to the vertical thermodynamic structure to identify gust potential. Microbursts (small-scale downdraft) can be produced where the low levels of the atmosphere are very dry (such as the western US), as well as in regions with very moist low-level conditions (such as the southeast US). Frequently, forecast and warning operations are complicated by the short lifetime of many storms in weak shear. The dry microburst is associated with a deep, dry adiabatic surface-based mixed layer with moisture confined to a layer in the middle troposphere (e.g., Wakimoto, 1985). On the other hand, wet microbursts develop in an environment with some dry air aloft and a deep surface-based moist layer, such that the difference between the equivalent potential temperature at the surface and middle troposphere is typically greater than 20 °C (Atkins and Wakimoto 1991). These two events represent the two processes discussed above. Usually, microbursts affect only small areas for a short period of time, but some have produced wind gusts in excess of 50 m s^{-1}.

In strong vertical wind shear environments (on the order of 15-20 m/s in the lowest 6 km) the prediction of damaging wind gusts becomes somewhat more complex. Nontornadic supercells can produce very strong winds. Frequently, this occurs when the shear in the lowest part of the atmosphere is not very strong or the LCL height is high. It also can occur when the storm-relative flow in mid-troposphere is very small (Brooks and Doswell, 1993). Damaging wind gusts have been related to bow-shaped convective elements, observed by radar, called bow echoes (Fujita, 1978; Przybylinski, 1995). Bow echoes may occur as isolated thunderstorm cells, or comprise elements of extensive squall lines. When a larger-scale bow echo or a series of bow echoes produce a succession of downbursts over a widespread region, this is termed a derecho (Johns and Hirt 1987; Lopez 2005).

Derechos were first studied in the warm season, but can occur at any time of year. The early studies identified specific synoptic environments with extreme instability near east-west oriented surface fronts on the southern edge of stronger westerly winds aloft and with tropospheric flow parallel to the front (e.g., Johns et al. 1990). When this occurs, bow echoes rapidly move along the front. The southward extent is usually limited by a strong inversion, preventing convection to develop. As a result, meridionally extensive squall lines don't form. Another environment associated with derechos, particularly outside of the warm season has been in association with energetic baroclinic waves. In this scenario, squall lines tend to form along or ahead of the cold front with a series of bow echoes embedded within the line. CAPE is usually much less in these situations, but the winds aloft are quite strong and the overall pattern bears a resemblance to more classic tornado outbreak patterns.

Evans and Doswell (2001) and Coniglio et al. (2004) have looked at the environments associated with derechos using proximity soundings. They found a variety of environmental shear patterns, although shear extending above the lowest few kilometers of the atmosphere appears to be important. The relationship between the environmental conditions and the strength of the forcing in quasi-geostrophic terms is complex. Weakly-forced situations tend to have very large values of CAPE (almost all greater than 1000 J kg^{-1}). Weakly-forced situations also have lower values of low level shear (mean of 11.4 m s^{-1} over the lowest 2.5 km compared to 18.3 m s^{-1} for the strongly forced cases).

2.4 Heavy rains

All but a very small number of thunderstorms produce rain. Thus, for the most part, our concern for the ingredients follows the basic ingredients for convection. However, we need to consider evaporation and the motion of the storm. Evaporation is obviously a negative for heavy precipitation. As a result, high relative humidity is important in the troposphere. Note, however, that nearly saturated environments are unlikely to have strong updrafts, so that a balance between strong updrafts, implying that the storm pulls in a significant amount of environmental moisture, and low evaporation, implying that the storm is efficient at producing rain at the ground, is needed.

In order to sustain high rain rates for a long enough period of time to produce high rainfall amounts, several different processes may go on. First, supercell thunderstorms sometimes have prodigious rain rates greater than 200 mm hr^{-1} (Smith et al., 2001). As a result, nearly stationary supercells can produce very large rain amounts quickly. Another way is for storms to "train", one following behind another, all dropping rain in nearly the same location. Many historically high rainfall events have followed this process (Doswell et al., 1996). In the case where a line of convection forms, perhaps along a synoptic boundary, and the tropospheric is parallel to the line, large rainfall amounts can occur at one location even if individual cells move rapidly. This complicates the forecast process as small errors in the forecast of storm motion can lead to incorrect forecasts. Monitoring boundaries and storm motion in relation to them can be helpful.

Frequently, as meteorologists, we are interested in flash floods, rather than simply heavy rain. Doswell et al. (1996) discuss the challenges associated with the difference between flash flood forecasting and heavy rain forecasting. Flash floods necessarily involve interactions with the ground. Antecedent precipitation, basin topography, and land surface conditions all can have significant impacts on the accumulation of water given an amount of precipitation.

3 Summary

An ingredients-based approach can be applied to any forecasting problem. It forces forecasters to recognize what they know about the atmosphere based on first principles and what they know that is based on their limited experience. It is possible that a combination of ingredients and processes that is common one location may not be as common in another place. Forecasters (and researchers) may believe they understand how a weather event occurs, but it could be biased because of the sample they have seen. As an example Trapp et al. (2004) produced a climatology of tornadoes from quasi-linear convective systems. In the southern plains of the US, tornadoes are dominated by supercell processes and the quasi-linear systems are much rarer. Farther east in the US, the quasi-linear system becomes more important. When the same set of ingredients and processes were put in place in the southern Plains that produce quasi-linear systems elsewhere, there is no reason to expect the systems to be incapable of producing tornadoes. Given the dominance of the supercell process in that region, it is not unreasonable that forecasters might not make an accurate forecast of the evolution, given their lack of familiarity with the situation. This could reflect a dependence on the "standard" model of tornadogenesis, rather than on the underlying ingredients and processes that underlie the model. In effect, this amounts to fitting the data to the model, rather than the other way around.

One of the perhaps unfortunate results of this is that it forces all of us to admit to, and confront, our ignorance. Forecasters don't have the option of not forecasting a weather event because they don't understand how it occurs. In turn, this defines important problems for research.

Acknowledgments

Discussions over many years with a number of people about the nature of forecasting have influenced my opinions on the subject. In particular, I would like to thank Chuck Doswell, Bob Maddox, and Steve Weiss for their ideas.

References

Atkins, N. T., and R. M. Wakimoto (1991). Wet microburst activity over the southeastern United States: Implications for forecasting. *Wea. Forecasting* 6: 470–482.

Brimelow, J C., G. W. Reuter and E. R. Poolman (2002). Modeling maximum hail size in Alberta thunderstorms. *Wea. Forecasting* 17: 1048–1062.

Brooks, H. E. (2005). Practical aspects of forecasting severe convection in the United States: Environmental conditions and initiation. This volume.

Brooks, H. E., A. R. Anderson, K. Riemann, I. Ebbers, and H. Flachs (2005). Climatological aspects of convective parameters from the NCAR/NCEP reanalysis. *Atmos. Res.*, in press.

Brooks, H. E., and J. P. Craven (2002). A database of proximity soundings for significant severe thunderstorms, 1957-1993. *Preprints*, 21st Conference on Severe Local Storms, San Antonio, Texas, American Meteorological Society, 639-642.

Brooks, H. E., and C. A. Doswell III (1993). Extreme winds in high-precipitation supercells. *Preprints*, 17th Conference on Severe Local Storms, Saint Louis, Missouri, American Meteorological Society, 173-177.

Brooks, H. E., J. W. Lee, and J. P. Craven (2003). The spatial distribution of severe thunderstorm and tornado environments from global reanalysis data. *Atmos. Res.* 67-68: 73-94

Brooks, H. E., and S. J. Weiss (1999). Severe local storms. *Proceedings of the WMO/UNESCO Sub-forum on Science and Technology in Support of Natural Disaster Reduction* WMO-No. 914, 12-31.

Coniglio, M. C., D. J. Stensrud, and M. B. Richman (2004). An observational study of derecho-producing convective systems. *Wea. Forecasting* 19: 320–337.

Craven, J. P., R. E. Jewell, and H. E. Brooks (2002). Comparison between observed convective cloud-base heights and lifting condensation level for two different lifted parcels. *Wea. Forecasting* 17: 885-890.

Doswell, C. A. III (1987). The distinction between large-scale and mesoscale contribution to severe convection: A case study example. *Wea. Forecasting* 2: 3-16.

Doswell, C. A. III, H. E. Brooks, and R. A. Maddox (1996). Flash-flood forecasting: An ingredients-based methodology. *Wea. Forecasting* 11: 360-381.

Evans, J. S., and C. A. Doswell III (2001). Examination of derecho environments using proximity soundings. *Wea. Forecasting* 16: 329–342.

Fujita, T. T. (1978). *Manual of downburst identification for project NIMROD*. Satellite and Mesometeorology Res. Paper No. 156, Dept. of Geophys. Sci., Univ. of Chicago, 104 pp.

Johns, R.H. (1993). Meteorological conditions associated with bow echo development in convective storms. *Wea. Forecasting* 8: 294-299.

Johns, R. H., and W.D. Hirt (1987) Derechos: Widespread convectively induced windstorms. *Wea. Forecasting* 2: 32-49.

López, J. M. (2005). A Mediterranean derecho: Catalonia (Spain), 17th August 2003. *Atmos. Res.*, submitted.

Markowski, P. (2005a). The concept of buoyancy and its application to deep moist convection. This volume.

Markowski, P. (2005b). Supercell thunderstorms. This volume.

Markowski, P. (2005c). Tornadoes and tornadogenesis. This volume.

Przybylinski, R. W. (1995). The bow echo: Observations, numerical simulations, and severe weather detection methods. *Wea. Forecasting* 10: 203-218.

Rasmussen, E. N., and D. O. Blanchard (1998). A baseline climatology of sounding-derived supercell and tornado forecast parameters. *Wea. Forecasting* 13: 1148–1164.

Smith, J. A., M. L. Baeck, Y. Zhang and C. A. Doswell III (2001). Extreme rainfall and flooding from supercell thunderstorms. *J. Hydrometeorology*, 2: 469–489.

Wakimoto, R.M. (1985). Forecasting dry microburst activity over the high plains. *Mon. Wea. Rev.* 113: 1131-1143.

Practical Aspects of Forecasting Severe Convection in the United States: Environmental Conditions and Initiation

Harold E. Brooks

NOAA/National Severe Storms Laboratory, Norman, OK, USA

Abstract. The first stage of forecasting convective weather involves forecasting the evolution of conditions that are favorable for the development of storms and their probable initiation. The scale of the forecasts are typically on the order of 100 km or larger and the lead time between the forecast and storms is 1-48 hours. In the United States, procedures have evolved so that the Storm Prediction Center of the National Weather Service has the responsibility for issuing these forecasts for the contiguous 48 states (the part of the US excluding Alaska and Hawaii.)

1 Background

In 1884, John Park Finley of the United States Army Signal Corps issued experimental forecasts for tornado occurrence in the United States. The forecasts were dichotomous forecasts (yes or no) for the occurrence of at least one tornado during a day in a set of districts around the US. The results of those forecasts ended up having greater impact on the forecast verification community (Murphy, 1996) than on the forecasting community as organizational changes curtailed the forecasts after only one season. Finley's vision of a central forecasting location was reborn in 1953 with the creation of a forecasting unit within the National Severe Storms Project (Corfidi, 1999). That unit has now become the Storm Prediction Center (SPC) and has a variety of forecasting responsibilities. The SPC's primary mission is to provide timely forecasts of severe thunderstorms and tornadoes, with lead times of up to three days. A variety of forecast products, some regularly scheduled and some issued as needed, are used to accomplish this mission. Before consideration of the kinds of information used to create these forecasts, I'll review the kinds of products that are issued and their relationship to the user community.

The underlying philosophy behind the approach to forecasting severe thunderstorms in the US has been referred as "ready-set-go" (Brooks and Weiss 1999). It is based on the results of sociological studies that indicate that, rather than take protective action, people tend to seek more information when first informed of a threat. The first products, issued on a regular basis up to three days in advance, are referred to as *Convective Outlooks*. They are issued once per day for the "Day Three" outlook (day after tomorrow), twice per day for the "Day Two" outlook (tomorrow), and five times per day for the "Day One" outlook, ending at 1200 UTC. The forecasts cover the so-called convective day, 1200 UTC to 1200 UTC, with the day starting at the minimum in the diurnal cycle of severe convection. The Day Three and Day Two outlooks are for the probability of severe thunderstorm (tornado, hail at least 19 mm in diameter, and/or wind gusts approximately 93 km h^{-1} or higher) occurring within approximately 40 km of any location, and the probability of a significant severe thunderstorm (tornado of at least F2 intensity on the Fujita scale, hail at least 50 mm in diameter, and/or wind gusts of 120 km h^{-1} or higher) (Fig. 1).

The graphical product also has a categorical equivalent to the probabilistic forecasts, with risk categories of High, Moderate, and Slight risk. The categorical products are similar to what has been issued historically. In addition to the graphical product, a text product describes the kinds of threats and the meteorological reasoning behind the forecast. The Day One products provide separate probabilities of the occurrence of tornadoes, hail, and wind gusts, but are similar in other ways to the Day Two and Day Three products. Although the general public may look at the outlooks, the primary audiences are local National Weather Service (NWS) offices, who have the responsibility to issue warnings as necessary for severe thunderstorms, and emergency management officials, who have the responsibility to prepare for events. The fact that the primary users are a relatively small group with a professional responsibility to be interested in the products means that the SPC forecasters can provide a fairly sophisticated discussion that varies from forecast to forecast.

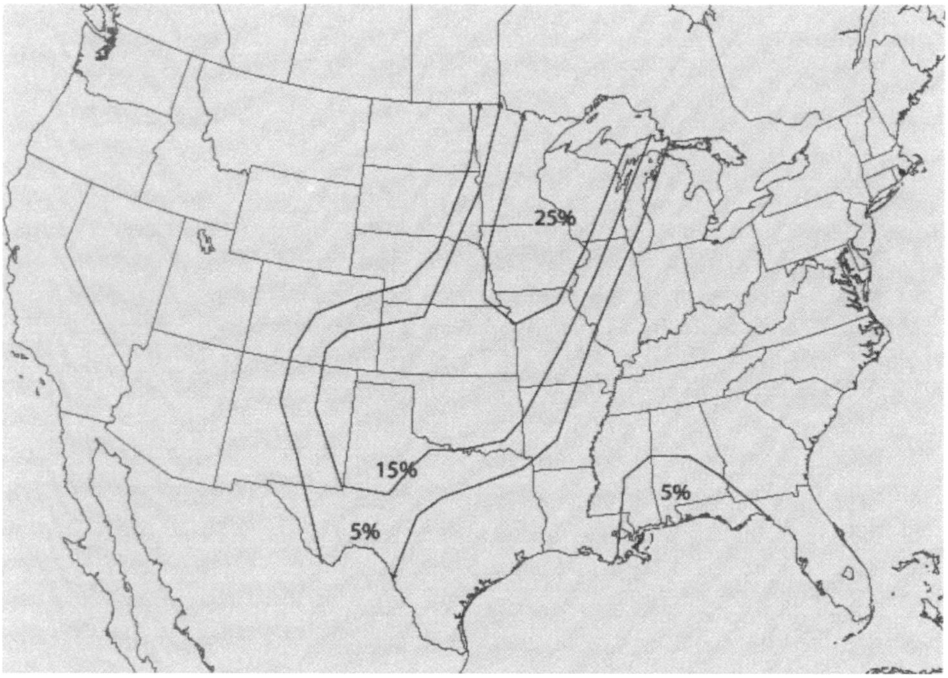

Figure 1: Sample Day Two convective outlook, valid starting 11 June 2005. Contours are probability of severe thunderstorm occurring within approximately 40 km of any location.

If the threat of severe thunderstorms or tornadoes becomes significant on a day, a *Watch* may be issued for the particular threat. Watches typically cover a time period of about 6 hours and an area of about 40,000 km^2. They are issued for tornadoes and for severe thunderstorms, indicating a lesser threat of tornadoes. Watches typically issued shortly before or after thunderstorms form. Watches are intended more for the public than outlooks are and, in particular, they are used by the news media to communicate with the public. As such, they tend to have a more structured format than the outlooks and provide similar information (nature of the threat, location and time frame of the watch, expected storm motion) in each watch (Fig. 2). The repetitive nature is intended to make it easier for less-sophisticated users to pick out the most important information.

URGENT - IMMEDIATE BROADCAST REQUESTED
TORNADO WATCH NUMBER 449
NWS STORM PREDICTION CENTER NORMAN OK
1130 AM CDT FRI JUN 10 2005

THE NWS STORM PREDICTION CENTER HAS ISSUED A
TORNADO WATCH FOR PORTIONS OF

SOUTHWEST KANSAS
NORTHWEST AND WEST CENTRAL OKLAHOMA INCLUDING THE PANHANDLE
THE TEXAS PANHANDLE

EFFECTIVE THIS FRIDAY MORNING AND EVENING FROM 1130 AM UNTIL 600 PM CDT.

TORNADOES...HAIL TO 2.5 INCHES IN DIAMETER...THUNDERSTORM WIND GUSTS TO 70 MPH...AND DANGEROUS LIGHTNING ARE POSSIBLE IN THESE AREAS.

THE TORNADO WATCH AREA IS APPROXIMATELY ALONG AND 85 STATUTE MILES EAST AND WEST OF A LINE FROM 15 MILES SOUTHEAST OF LUBBOCK TEXAS TO 35 MILES NORTHEAST OF LIBERAL KANSAS. FOR A COMPLETE DEPICTION OF THE WATCH SEE THE ASSOCIATED WATCH OUTLINE UPDATE (WOUS64 KWNS WOU9).

REMEMBER...A TORNADO WATCH MEANS CONDITIONS ARE FAVORABLE FOR TORNADOES AND SEVERE THUNDERSTORMS IN AND CLOSE TO THE WATCH AREA.
PERSONS IN THESE AREAS SHOULD BE ON THE LOOKOUT FOR THREATENING WEATHER CONDITIONS AND LISTEN FOR LATER STATEMENTS AND POSSIBLE WARNINGS.

DISCUSSION...AIR MASS CONTINUES TO TRY TO DESTABILIZE AHEAD OF APPROACHING MID/UPPER LEVEL TROUGH ACROSS EASTERN NM. CLOUD COVER IS LIMITING HEATING...BUT ASCENT FROM APPROACHING TROUGH IS INITIATING FURTHER STORM DEVELOPMENT. MESOANALYSIS SHOWS WEAKENING INHIBITION TO ALLOW ACTIVITY TO BECOME MORE SURFACE BASED WITH TIME...AND WITH SEVERAL BOUNDARIES ACROSS THE AREA AND FAVORABLE SHEAR...LOOKS LIKE MAIN THREATS WILL BE HAIL AND ISOLATED TORNADOES.

AVIATION...TORNADOES AND A FEW SEVERE THUNDERSTORMS WITH HAIL SURFACE AND ALOFT TO 2.5 INCHES. EXTREME TURBULENCE AND SURFACE WIND GUSTS TO 60 KNOTS. A FEW CUMULONIMBI WITH MAXIMUM TOPS TO 500. MEAN STORM MOTION VECTOR 22035.

Figure 2: Text of watch message associated with first watch issued for period covered by convective outlook in Fig. 1. "Discussion" and "Aviation" sections are not broadcast to the general public.

The last level of statement is the *Warning*. This indicates that a particular thunderstorm is producing severe weather or is about to. As with the watch product, warnings can either be for tornadoes or for severe thunderstorms. Warnings are issued by one of the approximately 120 local forecast offices in the US, each with forecast and warning responsibilities for areas on the order of 100,000 km^2. Warnings almost always are valid for one hour or less and typically cover a single county or portion of a county, an area in much of the country of about 2,500 km^2. They follow a standard format and include information on where the storm is, what information led to the warning (radar, spotter report of severe weather, etc.), where the storm is going, and conclude with a call to action recommending a specific response from the public. Warning procedures will be covered elsewhere (Brooks 2005b).

2 Meteorological Considerations

The focus here is on the meteorological information that goes into the development of convective outlook and watch products. In many senses, the procedures are similar, although the different time and space scales, as well as lead time, of the products means that forecasters can focus on more detail on the watch product.

At the heart of the forecast process is an ingredients-based approach to forecasting (Brooks 2005a). Based on our scientific understanding of what is required to produce a particular severe weather event, SPC forecasters consider the distribution of the ingredients and the processes that can change that distribution over the appropriate time scales. At one level of approximation, the standard procedures within the SPC amount to first identifying the potential for supercell thunderstorms, and then attempting to determine if they are likely to produce tornadoes, hail, and/or nontornadic wind gusts.

As discussed in Brooks (2005a) and Markowski (2005), the initial focus on supercells means that forecasters diagnose and predict the distributions of moisture, instability, lift, and organizing shear. After that, discriminating between the kinds of severe weather requires looking at the appropriate ingredients. SPC forecasters spend a considerable amount of time in the diagnosis stage of the forecast process. Analysis of synoptic upper air data on mandatory levels (925 hPa, 850 hPa, 700 hPa, 500 hPa, and either 300 hPa or 250 hPa) is done by hand each 12 hours. A full national surface analysis is carried out on surface data every three hours and regional analyses, focusing on areas of concern, are typically done on an hourly basis. Profiles of the thermodynamic and wind profiles from radiosonde ascents are considered in regions where convection is possible. There are few rigid requirements for the analyses, allowing individual forecasters freedom to highlight information of importance to the forecast problem of the day.

The analysis activities serve several purposes. First, it helps forecasters understand what is going on. Their connection with observed data helps them to be more aware of unusual (or suspect) observations. At a basic level, it helps them understand how the atmosphere is fitting together compared to the conceptual models they may have had for the day. Second, simple forecasts of changes from time period to time period can be made. In the short-term, extrapolation can be a useful forecast. Third, the analysis prepares them for interpreting numerical model forecasts. If the models are not reproducing observations, the forecasters will pick up on that fact more quickly.

Forecasters also look at a variety of numerical weather prediction (NWP) models. Although the models can be thought of as providing direct guidance about expected atmospheric development, the forecasters at the SPC also are involved in interpreting the models in terms of how the models relate to reality in practice. Frequently, in major severe weather events, the direct NWP guidance is not a particularly accurate forecast and, early in the forecast period, it is clear that significant differences exist between observed and predicted evolution (e.g., Edwards et al. 2002). Thus, many studies involving forecasters at the SPC and researchers at the National Severe Storms Laboratory (NSSL) have focused on maximizing the utility of the NWP forecasts. As examples, parameterized processes in the models don't necessarily produce output that reproduces structures seen in the atmosphere, whether they are correct on a particular day and location or not. Attempting to relate parameterized output with real atmospheric conditions, so that forecasters can interpret the models effectively has been an important effort over the last decade (e.g., Cortinas and Stensrud 1995). Baldwin et al. (2002) looked at NWP convective schemes and related them to real forecast problems. In that case, forecasters were able to tell that forecast vertical profiles of the atmosphere were wildly incorrect, but were unaware that some of the unusual structures were the result of the shortcomings of the convective schemes and that the fact that the models had produced those structures provided clues about how the model "thought" the atmosphere would evolve. Thompson (1998) looked at wind profiles associated with NWP forecasts of supercells. Stensrud et al. (1997) developed relationships between NWP output and observed severe thunderstorms. In these latter two papers, it is important to note that the work was based, in large part, on relationships derived from studies of the relationships between observed conditions and weather, but that the relationships between the NWP and weather were not necessarily the same as between observed conditions and weather. In a sense, this is a formalization of the process of forecasters learning to adapt to model errors and biases. More formally, it is an example of the notion of the meteorological covariate (Brown and Murphy 1996), where we want to have a relationship between a weather event and a fundamental variable, but instead we have to settle for a relationship between a weather event and another variable that we may observe or forecast better than the desired variable. The fundamental analysis problem then becomes the description of the relationship between the two variables.

A challenging problem for forecasters is the mismatch between the frequency of surface data and upper-air data. The most reliable upper-air data (the radiosondes) come in every 12 hours, but surface data are collected on an hourly or more frequent basis. Observations from commercial aircraft can help provide information about the upper-air situation, but the data tend to be collected near hub airports or in the upper troposphere. Given that severe weather forecasters are concerned with rapidly evolving conditions, the lack of upper air data can be a severe hindrance. In order to attempt to deal with the problem, a high-resolution time and space NWP model that uses all available observations is initialized every hour and run for 3-12 hours, depending on time of day, in order to give a best-guess estimate of the three dimensional structure of the atmosphere. Again, forecasters recognize that the estimate has problems at times, but it likely a better solution than keeping the upper air portion of an analysis constant while changing the surface data or using longer lead-time NWP products.

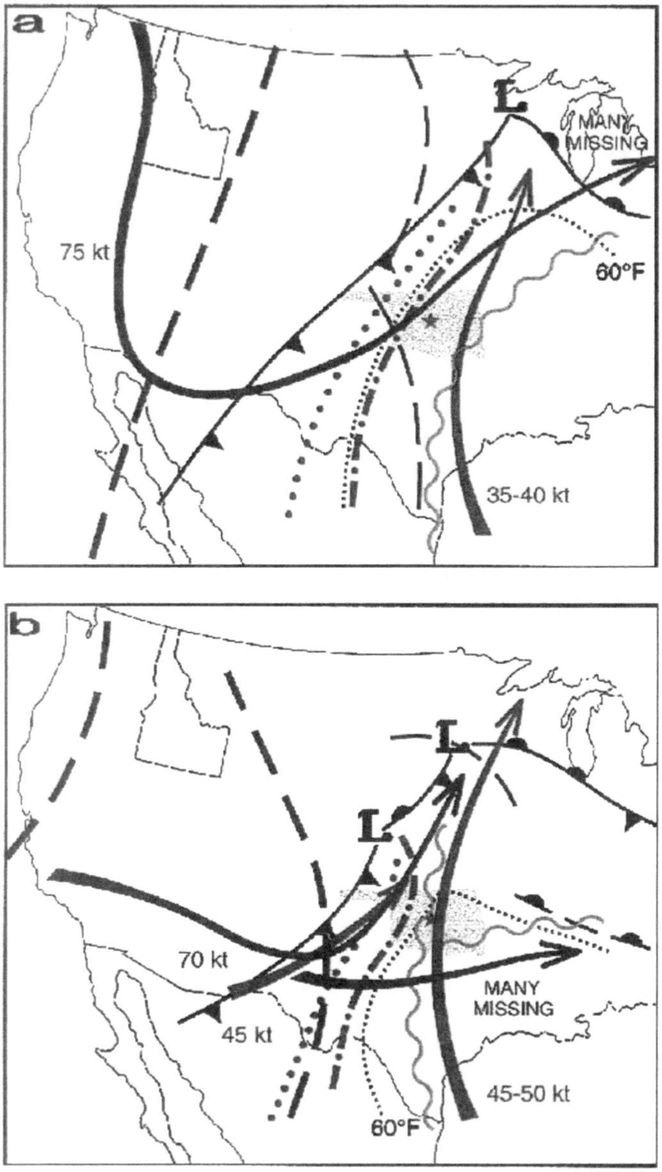

Fig. 3: Composite charts associated with a) 20 March 1948 and b) 25 March 1948 Tinker Air Force Base (shown by the purple star) tornadoes near Oklahoma City, Oklahoma. The shaded region is the state of Oklahoma. Features at 500 hPa in blue, 850 hPa in red and green, and surface frontal analyses and 60°F (15.5°C) dewpoint in black. Height troughs are dashed lines, solid arrows are jets, and dash-dot line is dryline location. The wavy green lines indicate moisture axes at 850 hPa and the red dotted lines the axes of highest temperatures at 850 hPa. Maximum observed wind speeds are indicated near the jets. Many winds were missing at 500 hPa, so jet values could be incorrect at that level. (Taken from Maddox and Crisp (1999).)

3 Logistical and Procedural Considerations

Clearly, the analysis requirements described in section 2 require considerable staffing. There are 4 or 5 forecasters on duty at all times in the SPC. Although they have specific responsibilities for putting out forecast products (e.g., the lead forecaster issues all watches and the outlook forecaster will put out the Day One outlook), there is reasonable flexibility in the procedures to the extent that, for instance, who analyzes which upper air charts may change from shift to shift. The goal is to come up with a common view of the evolution of the atmosphere amongst the group of forecasters on shift. Although the lead bears ultimate responsibility and may have to resolve conflicting views, in general, broad agreement is typically reached easily. One aspect of the problem is the need to simplify the large amount of information down to its critical features. One approach to this is via the use of composite charts, on which the most important features at each level of the atmosphere are noted. Composite charts for the March 1948 tornadoes at Tinker Air Force Base illustrate the effective use of this technique to highlight where important ingredients are coming together (Fig. 3). Although more detail will be needed to issue the necessary forecasts, the composite chart can highlight the most significant areas and focus attention for more detailed analysis. These charts can be produced from observed data or for forecasts at any lead time. Thus, a series of charts covering the appropriate forecast time can be an aid in producing a first guess at a forecast.

One problem that the SPC faces as an organization is the desire for forecast continuity, particularly in the outlooks. A non-meteorological consideration is that the forecasters do not want to have forecasts valid at the same time that "bounce" wildly from one solution to the next depending on issuance time. A major reason for this is that it is recognized that users may not always see the latest forecast and that forecasts continue may continue to be disseminated even after they have been replaced. Thus, avoidance of what might appear to be conflicting guidance is possible. As a result, even if evidence has emerged to suggest that the current forecast is inaccurate, changes may be limited by the desire to have the forecasts change smoothly. This is particularly true if a forecast has indicated a high probability of an event occurring. If new information suggests that the probabilities are much too high, the forecaster may not reduce the probabilities as much as they truly believe they should, in order to preserve continuity. This process has benefits beyond continuity. It builds cohesiveness in the unit between forecasters who don't work shifts together very often. It also limits the possibility of downplaying an event that was previously forecast as more likely and the follow-on negative repercussions of having been right about an event and then making the forecast wrong. Given that there is an asymmetric penalty function for bad forecasts (the costs of a missed event are almost certainly much higher than for a false alarm), this is a reasonable policy.

4 Beyond the Forecast

One aspect of the success of the SPC forecasters has been their involvement in research. Many SPC forecasters publish formal and informal manuscripts on a relatively frequent basis. In part, this is because the act of forecasting on a daily basis puts them on the edge of the state of the scientific understanding. Severe weather forecasters are likely to encounter events that are not well-understood by any meteorologist. As a result, they discover research problems on a regular basis. (Unfortunately, many of them come in the guise of a missed forecast!) The act of trying to understand what occurred helps improve the forecast at the next opportunity. Collocation with a research center (NSSL) has also provided the SPC with exposure to cutting-edge research via seminars and interpersonal communication. Researchers are encouraged to interact with the forecast staff and the proximity helps attract research staff that are interested in operational forecast problems. Thus, knock-on effects of the interest in research by the SPC staff and support from their management for research provides solutions to their forecast problems in the long run.

Another more formal activity that has proved successful and could, given the appropriate commitment of resources, be applied in other settings has been the forecast experiments sponsored jointly by the SPC and NSSL. These experiments are carried out for periods of roughly 6-10 weeks each year and have dealt with a variety of aspects of forecasting, from winter weather to hail size forecasting to the interpretation of NWP output to the use of extremely high-resolution NWP data. Forecasters who are not on duty, SPC supervisory personnel, NSSL researchers, and outside visitors, including academics, and forecasters from local offices, the private sector and foreign weather services, work together for a week at a time to study a narrow set of forecast problems. Through this exercise, new techniques are tested, modified, and adopted or dropped. It also encourages members of the research community to work on solving problems of importance to the SPC. It appears to have played a role in making forecasters more open to new ideas as well.

References

Baldwin, M. E., J. S. Kain and M. P. Kay (2002). Properties of the convection scheme in NCEP's Eta model that affect forecast sounding interpretation. *Wea. Forecasting*, 17: 1063–1079.

Brooks, H. E. (2005a). Ingredients-based forecasting. This volume.

Brooks, H. E. (2005b). Practical aspects of forecasting severe convection in the United States: Storm evolution and warning. This volume.

Brooks, H. E., and S. J. Weiss (1999). Severe local storms. *Proceedings of the WMO/UNESCO Sub-forum on Science and Technology in Support of Natural Disaster Reduction* WMO-No. 914, 12-31.

Brown, B. G., and A. H. Murphy (1996). Verification of aircraft icing forecasts: The use of standard measures and meteorological covariates. *Preprints*, 13th Conf. Probability and Statistics in the Atmospheric Sciences, San Francisco, California, USA, Amer. Meteorol. Soc., 251-252.

Corfidi, S. F. (1999). The birth and early years of the Storm Prediction Center. *Wea. Forecasting*, 14: 507–525

Cortinas, J. V. Jr., and D. J. Stensrud (1995). The importance of understanding mesoscale model parameterization schemes for weather forecasting. *Wea. Forecasting* 10: 716-740.

Edwards, R., S. F. Corfidi, R. L Thompson, J. S. Evans, J. P. Craven, J. P. Racy, D. W. McCarthy and M. D. Vescio (2002). Storm Prediction Center forecasting issues related to the 3 May 1999 tornado outbreak. *Wea. Forecasting*, 17: 544–558.

Maddox, R. A., and C. A. Crisp (1999). The Tinker AFB tornadoes of March 1948. *Wea. Forecasting*, 14: 492-499.

Markowski, P. (2005). Supercell thunderstorms. This volume.

Murphy, A. H. (1996). The Finley Affair: A signal event in forecast verification. *Wea. Forecasting*, 11: 3-20.

Stensrud, D. J., J. V. Cortinas Jr., and H. E. Brooks (1997). Discriminating between tornadic and nontornadic thunderstorms using mesoscale model output. *Wea. Forecasting* 12: 613- 632.

Thompson, R. L. (1998). Eta model storm-relative winds associated with tornadic and non-tornadic supercells. *Wea. Forecasting* 13: 125–137.

Practical Aspects of Forecasting Severe Convection in the United States: Storm Evolution and Warning

Harold E. Brooks

NOAA/National Severe Storms Laboratory, Norman, OK, USA

Abstract. In order to protect life and property, forecasts of severe convection are critical on short time and space scales (on the order of 1 hour or less and a few 10s of km or less). Accurate assessment of the environment and monitoring of high-resolution observational data, frequently focusing on radar-observed evolution, are essential in this process. In the United States, these short-term time and space scale forecasts are referred to as *warnings* and are prepared by local forecast offices of the National Weather Service, who have responsibility for forecasters on the order of 100,000 km^2.

1 Background

In the United States, very short-term [$\leq$O(1 h)], small-scale [$\leq$O(100 km)] forecasts of convection are the responsibility of approximately 120 local National Weather Service (NWS) Forecast Offices (FO). Their most important product is the warning, either for a tornado or a severe thunderstorm (in the US, hail of 19 mm or larger, convective wind gusts of 93 km h^{-1} or higher, or a tornado). The warnings typically cover one county or a part of a county and are in effect for an hour or less. In this paper, I'll consider aspects of the warning process, both meteorological and non-meteorological.

Larger scale forecasts are provided by the NWS Storm Prediction (SPC). The forecast process for the SPC is discussed in Brooks (2005). The convective outlook and watches from the SPC provide an important backdrop for FO operations. They highlight where large-scale or mesoscale conditions are favorable for severe thunderstorms to form. At the very least, they serve to draw the FO's attention to the potential for a problem. Typically, the watches are issued by the SPC after consultation with the FO about details of the boundaries.

At the warning stage, attention and responsibility falls almost entirely on the FO. The office's biggest tools are radar and information from spotters, although satellite and surface observations play a role.

2 Use of Radar

A national network of Doppler radars, the Weather Surveillance Radars-88 Doppler (WSR-88D) allows forecasters to look at storms in their region. Most FOs have a radar collocated with the office, but portions of several radars' coverage areas may include regions for which the office has warning responsibility.

Given the importance and ubiquity of the radars, a great emphasis has been placed on training in the interpretation and use of the data. During the deployment of the radars in the 1990s, all forecasters attended a training course to prepare them for the radar. Follow-up distance-learning

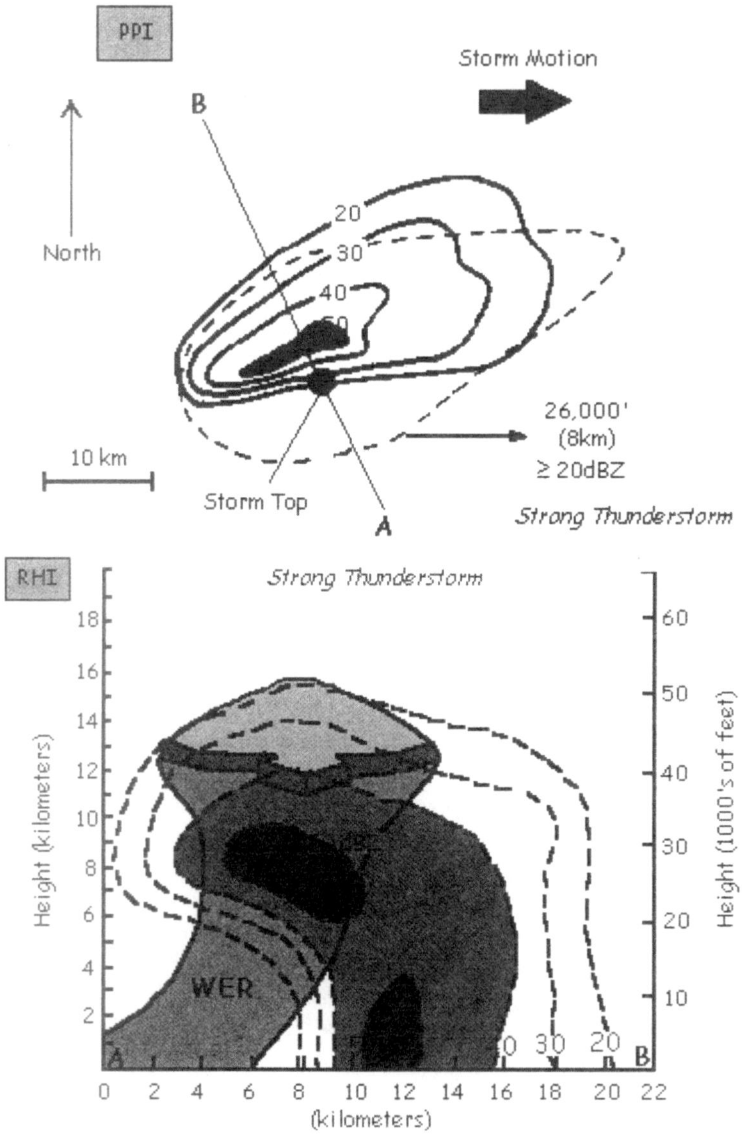

Figure 1: Illustration of principles of Lemon technique for a strong thunderstorm, taken from http://weather.cod.edu/ sirvatka/lt.html. On the top is a horizontal cross-section and the on the bottom is a vertical cross-section. Low-level reflectivity is shown by solid lines on the top figure and 8 km altitude 20 dBZ contour is dashed. In the bottom panel, reflectivity is shown in dashed contours with solid contours and arrow indicating inferred motion. Areas with greater than 50 dBZ reflectivity shown in darkest gray in both panels.

training materials were developed. In addition, the NWS Warning Decision Training Branch (WDTB) produces tornado warning guidance, advice for forecasters that includes information from recent improvements in scientific understanding of tornadogenesis.

Identification of reflectivity and velocity signatures associated with various severe thunderstorm structures is clearly important in the warning process. Just as in the larger scale of forecasts, determination of the potential of a storm to be a supercell is important. For reflectivity, the so-called Lemon technique (Lemon, 1977) is the basis for standard practice for identifying strong updrafts. Looking in the direction of storm motion, for a storm that moves to the right of the mean wind (the typical motion for Northern Hemisphere supercells that form when the storm-relative environmental winds veer with height), supercells tend to have notches in the reflectivity at low altitudes on the right side, associated with the inflow into the storm. At low levels, the right side of the echo tends to have a strong gradient from zero to high reflectivity. Above the inflow notch, there tends to a region of low reflectivity, the so-called weak echo region (WER), associated with the presence of the storm's strong updraft which does not allow enough time for precipitation particles to grow large enough to produce a significant echo. In some cases, the WER is completely surrounded by higher reflectivities forming a bounded weak echo region (BWER). Above the WER, some of the strongest reflectivities are found, a product of the large hail and heavy rain kept suspended by the storm updraft. The result is an area of low reflectivity directly under an area of higher reflectivity (Fig. 1).

At times, the rotation associated with the storm's mesocyclone can wrap precipitation around the back of the updraft, so that an appendage or "hook" of reflectivity forms to the back right side of the main echo. Although frequently associated with the presence of tornadoes, not all hook echoes produce tornadoes (Markowski, 2002). The relationship between the hook and tornado is not obvious.

Velocity data gives more direct indications of supercell structures by providing evidence for the rotation associated with the storm. In the northern hemisphere, areas of velocity away from the radar to the right and areas with velocity towards the radar are associated with cyclonic rotation. It is possible to have this velocity signature in pure shear flow, but other information, such as reflectivity or a time history of velocity, can typically help distinguish between rotation and shear.

Other distinctive radar signatures exist. Bow echoes are typically associated with strong rear-to-front surface flow at the apex of the bow. In addition, the cyclonic circulation on the left side of the apex (in the Northern Hemisphere) is frequently strong enough to produce damaging winds.

In general, surface winds can be estimated directly from the velocity measurements if the radar beam is sufficiently close to the surface of the earth. Unfortunately, operational considerations mean the lowest-elevation radar beam is rarely, if ever, located at 0° elevation. The WSR88-D radars are operated with the lowest elevation at 0.5°. Combined with the curvature of the earth, the result is that the center of the lowest beam is approximately 1 km above the ground at 60 km from the radar. Thus, events at the surface have to be inferred from observations aloft. The mid-altitude radial convergence (MARC) velocity signature, for instance, uses convergence a few km above the ground as a precursor for strong surface winds, associating the convergence with a downdraft that produces strong outflow (Schmocker and Przybylinski, 2005).

Precipitation can be estimated from the reflectivity measurements. A number of so-called "Z-R" (Z-radar reflectivity, R-rainfall) relationships exist that seek to relate rainfall to the reflectivity. There are numerous problems with such simple relations, including the estimation of the drop size distribution, contamination by hail, melting of frozen precipitation, beam blockage, attenuation, etc. Polarimetric procedures, which involve polarizing the radar beam, have shown promise to improve the estimates (Ryzhkov et al., 2005), but require the construction of new radars or the retrofitting of old radars.

The strong reflectivity associated with hail that makes it a problem for radar estimation can aid in the detection of hail. Very high reflectivities (~65 dBZ) are typically diagnostic for the presence of hail. Again, polarimetric measurements have shown great promise in discriminating hail from heavy rain (Zrnič et al. 2001). Falling hail tends to tumble, so stones look like spheres to a radar beam. The large raindrops associated with heavy rain tend to flatten out as they fall. As a result, raindrops look very large to a horizontal beam and much smaller to a vertical beam.

Large hail can also have an indirect signature in reflectivity data. Hail stones scatter the incident radar beams. For large stones, such that the wavelength of the radar beam is comparable to the size of the hailstone, some of the radiation is scattered away from the radar. Some of that scattered radiation can get reflected back off the earth, back into the storm, and then returned to the radar. Since the backscattered radiation takes longer to return to the radar receiver, it appears to be farther away from the radar than the hail core. Thus, a spike of reflectivity appears directly behind the radar echo, away from the radar. This signature is referred to as a three-body scatter spike (TBSS) (Lemon, 1998).

Algorithms have been developed that take radar data and process it to detect and/or predict the occurrence of weather events of interest. Case studies with relatively few events, and large climatological data sets both have been used in the development and validation of these algorithms. Mesocyclones, tornado vortex signatures, severe hail, damaging downbursts have all algorithms developed for the WSR-88D[1]. Although the spirit of the techniques can be applied elsewhere, the exact details may not transfer directly. Differences in the climatological frequency of some observed quantities would likely mean that any threshold-based efforts would have difficulty in translation to other regions. Thus, great care must be taken in transferring such techniques.

3 Use of Spotter Information

As mentioned, radar information is available near the ground for only a small portion of the radar coverage area. Thus, we have very few observations in a region where important processes are taking place. In addition, some physically-relevant features do not lend themselves well to automated observing systems. As a result, observations from human beings can play an important role in providing information to the forecaster in severe thunderstorm situations. The nature and organization of cloud-base rotation and the relationship of wrapping rain curtains to low-level cloud structures are examples of things that can be observed and interpreted by trained human observers. In much of the US, members of public safety organizations and even the general public receive training as volunteer weather observers ("spotters") for severe weather situations. This training is typically organized by the Warning Coordination Meteorologist (WCM) of the local FO as part of their public outreach efforts. In many communities, the network of spotters is activated by local public safety or amateur radio organizations, when

[1] See Vol. 13, No. 2 (June 1998) of *Weather and Forecasting* for an almost complete description of the algorithms at that time.

requested by the FO. The activating organization then may act as a filter on the observations, combining observations of the same event from different locations, and filtering bad reports out. In turn, the FO can provide detailed information to the local public safety group. Another side benefit is the development of a well-informed user community (the trainee, families, and friends) that understands weather information and is prepared to take safety measures as needed.

4 Warning Decisions

Andra et al. (2002) have described the challenges associated with operational decision making for warnings. In part, these challenges exist because of the need to make decisions under extreme time pressure with uncertain information. Klein (1999) discusses successful approaches to these kinds of decision problems in a variety of fields outside of meteorology. In general, successful decision makers are able to assess the situation quickly, using an internal "library" of cases to which to compare the current one. They isolate the most important features of the situation and then quickly come to a solution of expected behavior and a strategy to deal with it. Klein (1999) points out that they may not find the "best" solution, but that they will find a good solution quickly, which allows for more time for action. The question of whether the "quick, good decision" is better than the "slow, better solution" is related to the third kind of forecast goodness identified by Murphy (1993), namely value. Obviously, a perfect tornado warning, perfect in the sense that it describes where a tornado will be accurately and precisely, with a few seconds lead time is likely to be preferred by few users to a less accurate and precise forecast with 15 minutes lead time. At the other extreme, a forecast with no precision and a very long lead time, such as the climatological frequency of the event, provides little value for users for a particular day's events. Brooks (2004) showed that, overall nationally, tornado warning performance can be modelled using signal detection theory. This implies it might be possible to estimate the effects of various kinds of changes on different measures of warning performance.

Forecasters are required to take into account many factors, including their expectations for the day, the history and current state of the observations, the location of the storm compared to sensitive areas, and the quality of the information available, in making warning decisions. Andra et al. (2002) point out the desirability of having someone in the FO who is not responsible for issuing warnings monitoring the situation to make sure storms or events don't slip past the system unwarned due to a lack of "situational awareness." Another aspect of improving the value of the warnings is preparing the user community (public safety and the general public) to know how to respond to warnings. A perfect forecast, in terms of accuracy and precision, will be of little value, if no one in the community knows what to do in the face of an event.

5 Summary

Warnings are an important part of any weather service. In the US, the quality of warnings has improved over the years. Brooks (2004) showed that the overall system of tornado warnings can be modelled statistically. Components of the improvement can't be broken out, but the overall improvement can be seen (Fig. 2). An important result of the work was that for a climatologically-rare event, if the cost of a missed detection is much higher than the cost of a false alarm (as is typically the case), we have to expect that most of the warnings will not have an associated event, given the current state of the science. This has important implications for the setting of goals for forecast performance.

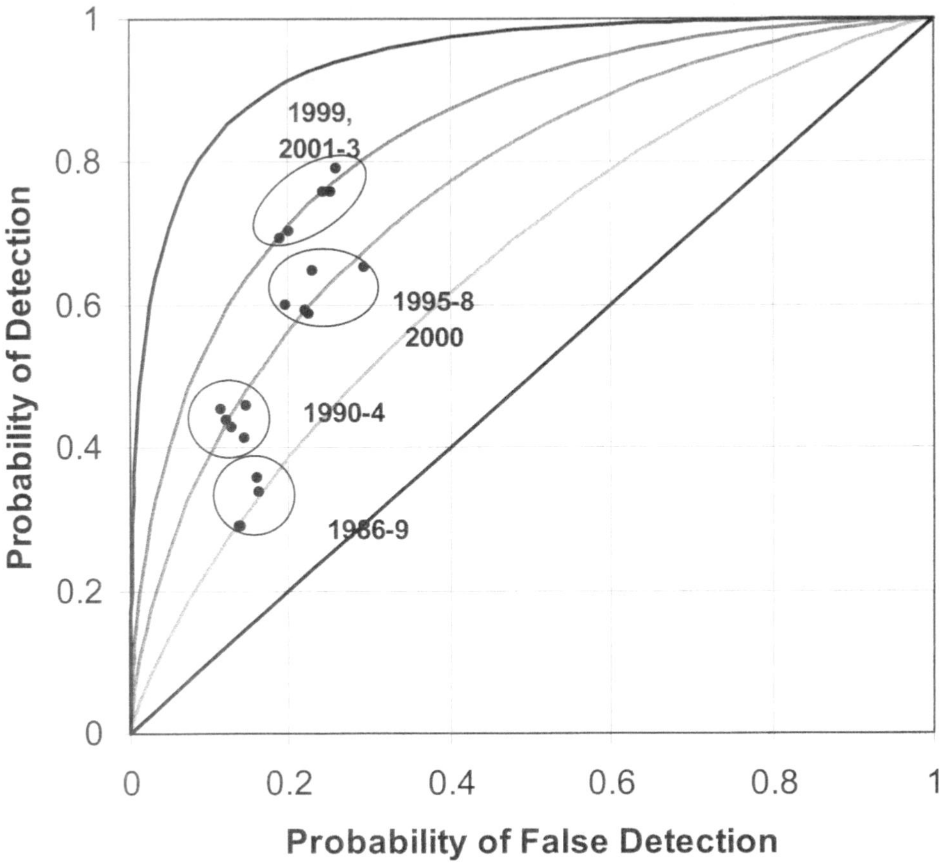

Figure 2: Relative operating characteristics diagram for national tornado warning performance in the US by year from 1986-2004. For each year, the probability of detection (tornadoes with warnings divided by total tornadoes) is plotted versus the probability of false detection (warnings without tornadoes divided by estimated number of events for which a warning decision had to be made). Adapted and updated from Brooks (2004).

Simmons and Sutter (2005) adapted econometric models to estimate the impact of the WSR-88D system on deaths and injuries from tornadoes. They took a number of input variables, including the intensity of the tornado, population and wealth of the county, time of day, and estimated that the WSR-88D system has reduced death tolls by approximately 45%, or roughly 40 deaths. Using standard estimates for the value of a statistical life in the US, this works out to an economic benefit of approximately $300,000,000 per year. It is important to note that the components of the radar (technology, training, human interpretation, etc.) cannot be separated out in this analysis.

Thus, it appears that warnings have improved and that tangible benefits can be measured. The process that led to this point has been going on since the early 1950s and represents a considerable amount of effort and institutional commitment, both from the operational sector to provide the warnings and from the research community to provide the information and tools that allow the forecasters to issue the warnings.

References

Andra, D. L. Jr., E. M. Quoetone, and W. F. Bunting (2002). Warning decision making: The relative roles of conceptual models, technology, strategy, and forecaster expertise on 3 May 1999. *Wea. Forecasting* 17: 559–566.

Brooks, H. E. (2004). Tornado warning performance in the past and future: A perspective from signal detection theory. *Bull. Amer. Meteor. Soc.* 85: 837-843.

Brooks, H. E. (2005). Practical aspects of forecasting severe convection in the United States: Environmental conditions and initiation. This volume.

Klein, G. (1999). *Sources of Power: How People Make Decisions.* MIT Press. 352 pp.

Lemon, L. R. (1977). Severe storm evolution: its use in a new technique for radar warnings. *Preprints*, 10th Conference on Severe Local Storms, Boston, MA, Amer. Meteor. Soc.,77-80.

Lemon, L. R. (1998). The radar "three-body scatter spike": An operational large-hail signature. *Wea Forecasting* 13: 327–340.

Markowski, P. M. (2002). Hook echoes and rear-flank downdrafts: A review. *Mon. Wea. Rev.* 130: 852–876.

Murphy, A. H. (1993). What is a 'good' forecast? An essay on the nature of goodness in weather forecasting. *Wea. Forecasting* 8: 281-293.

Ryzhkov, A. V., S. E. Giangrande, and T. J. Schuur (2005). Rainfall estimation with a polarimetric prototype of WSR-88D. *J. App. Meteor.* 44:502–515.

Schmocker, G., and R. Przybylinski (2005). Mid-altitude radial convergence (MARC) velocity signature. URL: http://www.crh.noaa.gov/lsx/science/marc.htm.

Simmons, K. M., and D. Sutter (2005). WSR-88D radar, tornado warnings and tornado casualties. *Wea. Forecasting* 20: in press.

Zrnič, D. S., A. Ryzhkov, J. Straka, Y. Liu, and J. Vivekanandan (2001). Testing a procedure for automatic classification of hydrometeor types. *J. Atmos. Ocean. Tech.* 18: 892-913.

General considerations on the operational forecasts of severe convective events: from medium to short range

Dario B. Giaiotti and Fulvio Stel
ARPA FVG OSMER, Regional Meteorological Observatory, Visco (UD), ITALY

Abstract In this lecture we will deal with the general aspects of an operational forecast of convective severe weather in the medium and short range, that is from $72^{\mathrm{h}}$to 24^{h} ahead the occurrence of the severe weather event. The attention will be focused on the information available to the forecaster, their reliability and their use. The role of the numerical model outputs generally available in the daily operational forecast activity are described and their limits are stressed. It is shown how the subjective contribution of the forecaster integrates the model outputs information. The main elements that characterize a severe weather occurrence are schematically described and their identification is explored by means of the useful information available at the medium and short range.

1 Difficulties in forecasting convective severe weather phenomena

Nowadays the weather forecasting activity, especially the operational one, is an efficient blend of objective techniques, objective evaluations and subjective choices.

Numerical models generate objective simulations which aim is that to describe the future state of the atmosphere by means of the numerical solution (Hutter and Jóhnk, 2004) of a basic closed set of differential equations starting from observed initial conditions (Gill, 1982; Holton, 1972; Batchelor, 1994). The quality of the numerical models outputs has increased since the first attempts, in the mid of the XXth century, and today global models are reliable in forecasting the synoptic features of the troposphere up to four days in advance, especially for mid latitudes. In many cases the quality of the predicted fields is still good a few days beyond the 4th day, but the chaotic behavior of the atmosphere limits a lot the use of those outputs, in particular when they are an element of the chain that leads to the local severe weather forecasts.

One of the principal limits of the global models outputs is their low spatial resolution, see figure 1, in comparison with the typical scales of the convective cells and the areas where those born and develop. Generally the forecaster deals with global models fields with an horizontal resolution of about 0.5° that means about 50 km. The vertical resolution is almost good for the forecasting purposes. For this reason, limited area models, shortly called LAMs, are run with a larger resolution in horizontal (about 15 km),

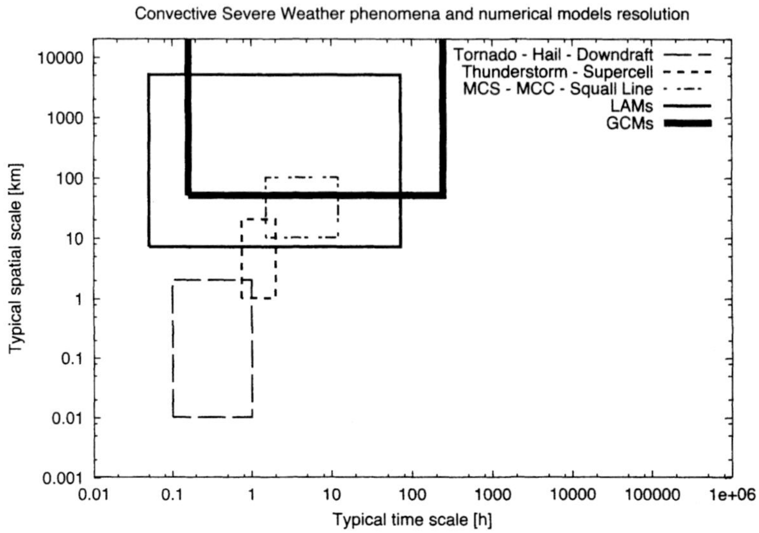

Figure 1. Schematic comparison between time and horizontal typical scales of some of the most important severe weather phenomena, that are produced by deep atmospheric convection, and the space-time domain covered by the numerical models simulations. It is shown that global circulation models have an horizontal spatial resolution not suitable for resolving most of the convective severe weather phenomena, while limited area models could in principle deal with the larger ones, but not with those strictly related to the cloud or subcloud scales.

but their outputs are generally affected by errors generated by a significant poorness in the initial conditions representing the real state of the atmosphere at the resolution they are set to resolve the future states. Usually LAMs are simply nested in the global models outputs and only some of them assimilate local meteorological measurements. Furthermore the nesting process propagates the uncertainties of the global model outputs in the limited area simulation. There are many LAMs running operationally and their products are often disseminated in combination with those of their parent global model.

According to these facts, it is clear that the model outputs operationally available to the forecaster are not able to reproduce all the features of the atmospheric deep convection. Because of that, the contribution of the forecaster is pivotal in the forecasting process. The forecaster integrates the numerical model outputs with other information that are not included in the models. These information are mainly data at the mesoscale and the forecaster experience in identifying the convective severe weather scenario over the interested domain. In many cases mesoscale information are summarized in indexes or simple objective downscaling tools that attempt to go beyond the limits of the classical numerical simulations. When the forecast has a very short lead time, less than 3-6 hours, usually it is called nowcasting and it is mainly based on real data, that is weather station data, radiosounding profiles, radar and satellite imagery, lightning records, Earth electric field records and so on.

In these lectures you will find a description of the path that a forecaster operationally follows to evaluate the possible occurrence of severe weather events at the mid latitudes over a terrain with a quite complex orography and the presence of a shallow sea. The approach is typical of the daily operational forecast, that attack the problem about 72 or 48 hours ahead, it follows the evolution of the troposphere and it updates the forecast getting closer to the event according to the new information available. The process is naturally concluded with the nowcasting.

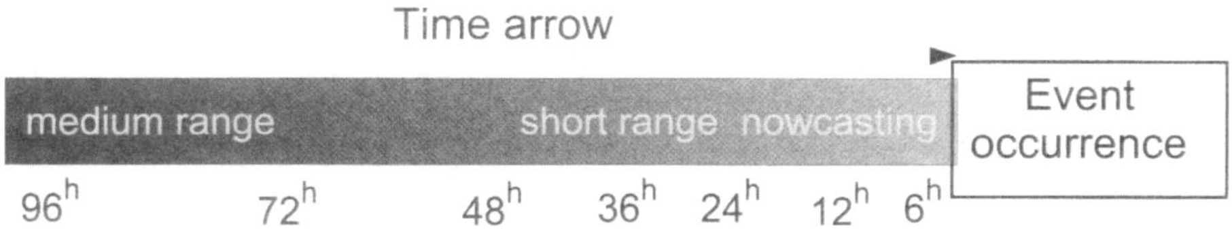

Figure 2. For the purposes of this course, the medium range and short range forecasts are defined as the forecasts that consider the posibility of a convective severe weather episode at least 24 hours before its occurrence. Nowcasting follows and it becomes dominat from 3-6 hours ahead to event onset. This definition is based on the fact that the relevant and reliable information for the evaluation of a severe convective episode, at mid latitudes, are mainly available about less than 24 hours ahead.

Whenever possible, besides the explanation on how to get and to analyze the suitable information required in the forecasting process, links between theoretical aspects of the deep atmospheric convection and the operational needs will be stressed. Furthermore the development of objective tools helping the forecaster in the operational activity are going to be considered and references to already published works will be given.

Even if the approach to the operational forecast of convective severe weather phenomena is not unique, it is hoped that these pages could help young researchers and forecasters in finding the common relevant positive aspects and weak points of the approach they are going to adopt in identifying in advance the risks associated with atmospheric deep convection.

2 The basic ingredients of the convective severe weather

In forecasting convective severe weather events it is important to keep in mind the three main elements which are necessary conditions for the severe weather occurrence.

- **The tropospheric instability.**
- **The convection trigger.**
- **The boundary conditions of severe weather occurrence.**

To issue a successful weather forecast in case of convective severe weather and to avoid false alarms all these three elements have to be properly forecasted. Here we give just a brief description of these basic ingredients of the severe weather due to convection and in the next sections each of them will be considered in detail recalling some of the theoretical aspects useful for their definition, furthermore their practical evaluation in the frame of operational weather forecasts is going to be stressed.

The tropospheric instability can be regarded as the inclination of the troposphere to mix along the vertical by means of the transport of air masses from the bottom to the top of a tropospheric layer and vice versa. In many cases, especially those manifesting severe weather episodes, the whole tropospheric thickness is interested by mass transport. We can consider the instability as the fuel that the convection will use to maintain the vertical motions of the air. The instability of the troposphere has been studied since the beginning of the atmospheric physics and nowadays its theory is almost well defined and, even if there are still details to be fixed, it gives a reliable quantitative description of the mechanism responsible for its onset and evolution.

The convection trigger is a mechanical action that displaces air masses along the vertical as far as they start to move autonomously under the buoyancy force that characterizes the unstable environments. Following the similarity done between instability and fuel, the convection triggers can be regarded as the match that lights the fuel. There are several kinds of trigger for convection, many of them act on the air masses closest to the ground. The understanding of the convection triggers is quite good for some of them, but there are still some hypotheses of triggering actions that have been not deeply explored by means of an objective approach. In many cases the convection triggers rely on specific features of the domain over which the convection takes place, they are identified by the forecasters through their experience and they have been tested qualitatively only.

The boundary conditions of convective severe weather are all the environmental conditions that make the difference between a non severe convective storm and the manifestation of a severe convective phenomena. Continuing the similarity, the boundary conditions can be considered the comburent which added to the fuel produces an explosive mixture. Since the risks associated to the severe convection are many and of different type, i.e. strong winds, tornadoes, hail, heavy precipitation and intense electrical activity, then it is likely that an overall and unitary model for the special conditions leading tropospheric convection to a severe weather episode should be in principle quite complex. So far, important aspects of the boundary conditions of specific severe weather manifestations have been deeply explored, for example the environments prone to the tornado formation or the occurrence of large hailstones, anyway, also for these well studied phenomena there is not any conclusive objective evidence that has produced a universal quantitative tool usable in operational weather forecast. Nevertheless many ideas on boundary conditions are applied in operational forecasts, both quantitatively and qualitatively.

It is well known that explosive phenomena need the contribution of all the three elements: fuel, comburent and ignition, to take place. The same occurs for the atmosphere to produce convective severe weather events. As in the case of classical explosions, for the atmosphere, the maximum effect is obtained with a proper combination of those elements. Unfortunately so far we have not produced a valid conceptual model, completely successful, that could be used for daily operational forecast, that is with a negligible number of false alarms and missing events.

2.1 Some general considerations on convective severe weather.

Here, some useful general consideration are presented. These consideration are meant for allowing the reader to better evaluate the objectives of the operational convective severe weather forecasts and their basic limits. They are presented in a schematic form to stress they are summarizing experiences produced by operational forecasting activity, case studies and deep theoretical research works.

Observation 1. Deep tropospheric convection is a necessary condition for the occurrence of convective severe weather phenomena.

Observation 2. Not all deep convective episodes are associated with convective severe weather phenomena.

Observation 3. As a consequence of the above observations, to forecast deep convection is not equals to forecast the risks associated with convective severe weather phenomena. So, forecasting convective severe weather events is in principle more difficult than forecasting the occurrence of deep convection. Identifying tropospheric deep convection with severe weather phenomena will produce a larger amount of false alarms.

The main severe weather phenomena connected with deep atmospheric convection are shortly listed below. Some threshold values are included, but they are not universally valid. In fact each area interested by deep convection has its own typical thresholds for the identification of the severe weather occurrence, because of the different response of the environment, the human activities and the local society to the occurrence of a strong weather phenomenon.

- **Strong winds.** This class includes all the phenomena that cause strong momentum transfer from the atmosphere to the surrounding environment by means of winds. Strong downdrafts and tornadoes are the main responsible phenomena. For the aviation purposes large horizontal wind shears and intense updrafts are included in this class.

- **Hydrometeors.** Large hailstones, with diameters conventionally larger than 2 cm, heavy rainfalls and snowfalls belong to this class. Rainfall threshold is the most area-sensitive parameter. For example in the Northeast of Italy values of 50 mm/h are not rare even for not severe weather events.

- **Electrometeors.** Cloud to ground lightning with an high spatial density and time frequency and large current intensity. Cloud to cloud strikes are also important especially for flying objects like planes, balloons, gliders etc. No thresholds for those are usually set. Earth electric field values, usually several thousand of Volts per meter are considered threatening human activities at the ground an on fly.

3 Numerical models outpus from 72 to 36 hours ahead the severe weather occurrence.

3.1 Forecasting the instability

Instability of the troposphere can be defined considering an adiabatic virtual displacement of an air parcel from an altitude to another, that is moving the parcel adiabatically along the vertical. According to the decrease of the atmospheric pressure with the height the moving parcel experiences an expansion if it is forced upward, while a compression if it is moved downward. Considering conventionally the upward displacement only, but the definition of instability can be done using downward displacements as well, if the parcel at the new position experiences an upward net force (buoyancy), then the tropospheric layer between the initial and final position of the parcel is considered unstable. On the other hand, if on the parcel acts a downward net force, attempting to restore the parcel to its original position, then the tropospheric layer is considered stable. In case of no resulting forces acting on the parcel, the neutral condition is defined and the parcel does not move from the final position. Since the displacement is considered adiabatic and the pressure of the parcel always equals the environmental one, then the buoyancy force acting on the parcel is:

$$F = \rho_p \frac{d^2 z}{dt^2} = (\rho_e - \rho_p) g \tag{3.1}$$

ρ_p is the parcel density, ρ_e is the environment density, g is the gravity acceleration, z is the vertical coordinate, positive upwards, t is the time.

This considerations, in combination with the ideal gas law:

$$p = \rho R T \tag{3.2}$$

(p is the gas pressure, ρ is the gas density, R is the gas constant and T is the gas temperature), leads to the expression of the acceleration for the parcel in terms of the difference between the temperatures of the air in the parcel (T_p) and that of the surrounding environment (T_e).

$$\frac{d^2 z}{dt^2} = \frac{(\rho_e - \rho_p)}{\rho_p} g = \frac{(T_p - T_e)}{T_e} g \tag{3.3}$$

So the parcel experiences an upward acceleration if $T_p > T_e$ and a downward acceleration if $T_p < T_e$. In case of $T_p = T_e$ the parcel is in the neutral condition.

Recalling the definition of potential temperature (Θ), that is the temperature of a dry air volume when it is brought adiabatically at the reference pressure (p_0) set conventionally to $1000 hPa$:

$$\Theta = T \left(\frac{p_0}{p}\right)^{\frac{R}{C_p}} \tag{3.4}$$

(all the variables here involved have already been defined in equation (3.2), while C_p is the specific heat for dry air at constant pressure ($\approx 1004\ J\,kg^{-1} K^{-1}$) and R is the gas constant for dry air ($\approx 287\ J\,kg^{-1}K^{-1}$), the equation (3.3) can be rewritten as a function of potential temperatures as follows:

$$\frac{d^2z}{dt^2} = \frac{\Theta_p - \Theta_e}{\Theta_e} g \tag{3.5}$$

This form of the buoyancy acceleration equation gives the possibility to have an estimate of the instability of the future tropospheric environment by means of a quick glance to the temperature fields available from numerical models outputs. In fact the thermal vertical profile of the troposphere in hydrostatic equilibrium, is characterized by the adiabatic expansion of the tropospheric gas only, that produce the consequent decrease of temperature. This fact is described in terms of potential temperature saying that the potential temperature is constant ($\Theta_e = constant$) along the vertical.

So from equation (3.5), if a parcel of air close to the Earth surface is displaced upward and along its path it finds an environmental potential temperature lower than itself, that is the environmental potential temperature decreases with height, then there will be instability. According to this considerations, for forecasting purposes, it is important to verify if the environmental potential temperature is decreasing rising with the altitude. The forecaster can quickly and easily evaluate this from model outputs using the thermal field of at least two isobaric levels.

In case of dry air, the decrease of in situ temperature is constant with height and depends upon only two important features of our planet: the first is the gravity acceleration (g) that is consequence mainly of the solid mass and shape of the Earth. The second one is related to the composition of the mixture of gases composing the dry troposphere, that is the specific heat at constant pressure C_p. This typical in situ temperature decrease is called lapse rate (γ_d) and it is defined as follows:

$$\gamma_d = -\frac{\partial T}{\partial z} = \frac{g}{C_p} \approx 10^{-2} K m^{-1} = 10 K km^{-1} \tag{3.6}$$

If the moisture of the air is considered, then the moist adiabatic lapse rate (γ_m) is computable, see Dutton (1995) and ? for more details, and its global climatological value is:

$$\gamma_m = (7 \pm 3) 10^{-3}\ K m^{-1} = (7 \pm 3)\ K km^{-1} \tag{3.7}$$

At mid latitudes, $7 K km^{-1}$ is a typical value representing the γ_m. So looking at the overall lapse rate, lets say the difference between the ground level temperature and the corresponding $500 hPa$ value, if the result is larger that the expected difference computed by means of the γ_m, then it is expected that instability will occur.

$$T_G - T_{500hPa} \geq \gamma_m \cdot Z_{500hPa} \tag{3.8}$$

T_G is the best forecast of the temperature for the tropospheric levels close to the ground, T_{500hPa} is the $500hPa$ temperature forecasted by the model and Z_{500hPa} represents the geopotential height of the $500hPa$ isobaric surface.

It is worth to note that the forecaster works on the environmental lapse rate and there is not any reference to the properties of the air parcel that will be buoyant. This

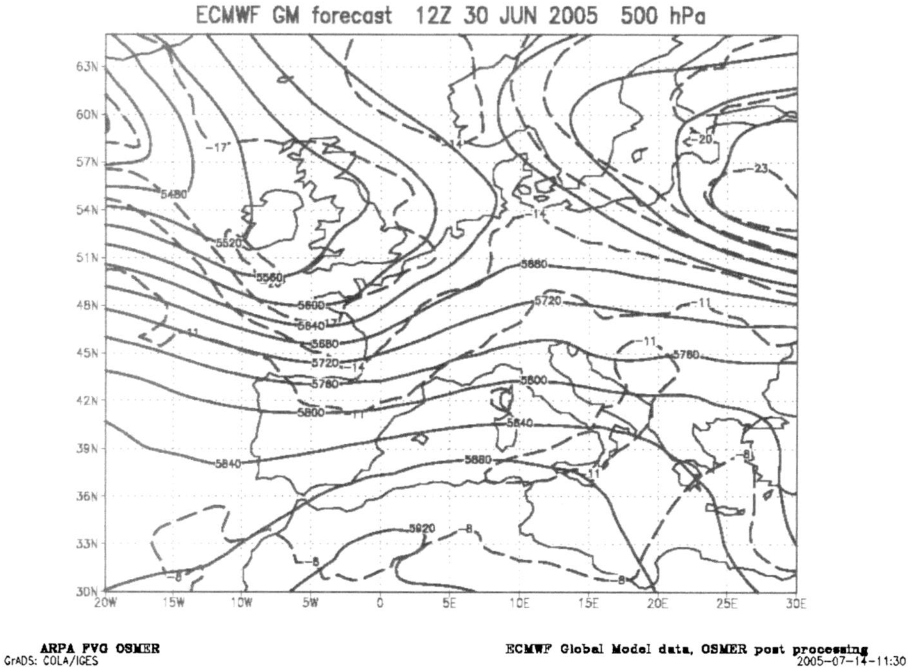

Figure 3. Example of a stable environment over the Po Valley (Northern Italy). The figure shows the ECMWF +48h forecast $500hPa$ temperature field in dashed lines and the corresponding geopotential height in black solid lines. The previous days were very hot over the plain with maximum temperatures almost higher than $30°C$, reaching peaks of $35°C$. The forecaster expected again maximum ground temperatures closer to or a bit greater than $30°C$ for the next two days over the Po Valley. Anyway aloft, the $500hPa$ temperature filed is characterized by a weak geopotential ridge with temperatures of about $-9°C$ and a warmer spot over the western Mediterranean, that is moving towards the Northern Italy. In this case the overall lapse rate is estimated to be $30°C-(-9°C) \approx 39°C$. The resulting overall moist adiabatic lapse rate produced by the product of γ_m by the $500hPa$ geopotential height, $5700m - 5800m$, results $\approx 40°C$. This suggests the forecaster that the environment is not particularly prone to the instability, furthermore the forecaster considers that the real lapse rate will decrease in the mid and upper troposphere because of the advection of warmer air due to the eastward motion of the warmer spot. The resulting forecast is that deep atmospheric convection is not likely to occur, that is one of the LSW ingredients is lacking, so no severe weather is expected for the central hours of the June, 30.

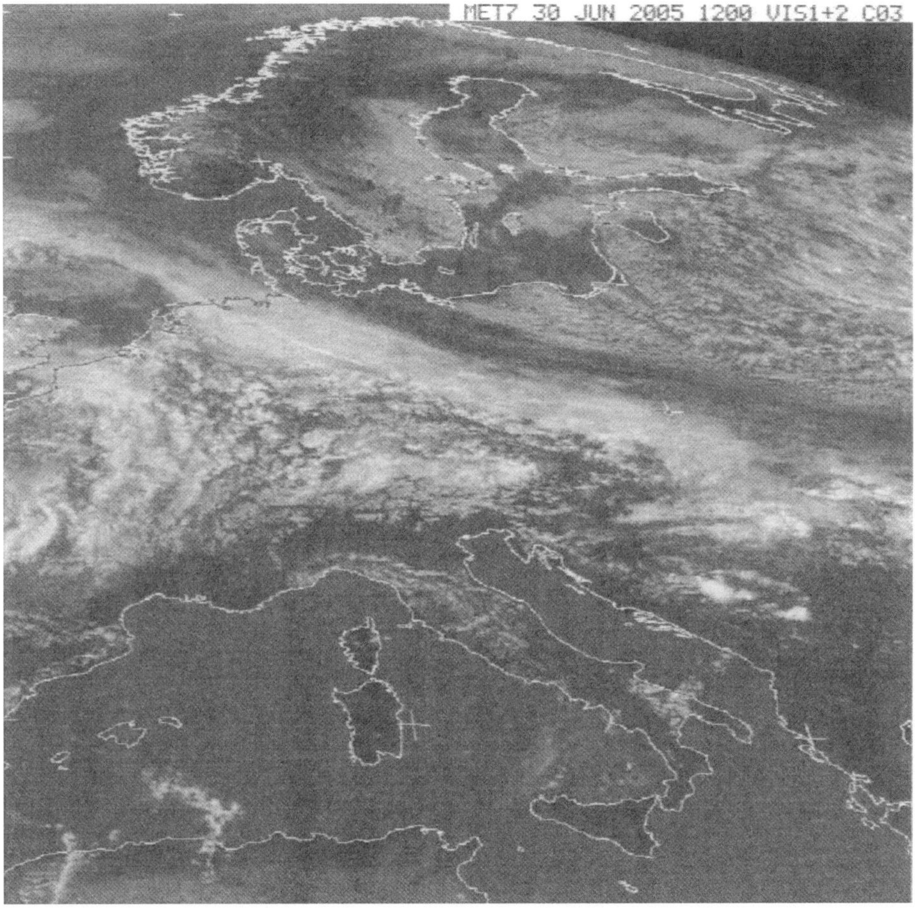

Figure 4. METEOSAT visible channel image at 12 UTC of June 30, 2005. There is no deep convection over the Po Valley. (copyright 2005 EUMETSAT)

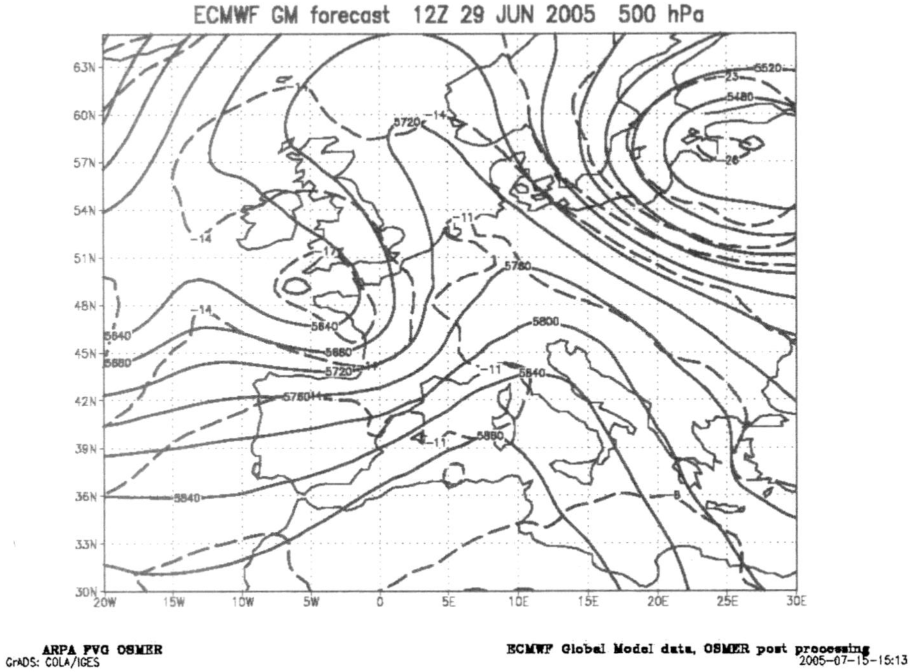

Figure 5. Example of an unstable environment over the western Europe and the North-east of Italy. The figure shows the ECMWF +48h forecast $500hPa$ temperature field in color and the corresponding geopotential height in black solid lines. Focusing the attention on the northeastern Italy, the forecaster expected maximum ground temperatures closer to $33°C$ for the next two days over the Po Valley. Aloft, the $500hPa$ temperature filed is characterized by a geopotential ridge with temperatures of about $-10°C$. The forecasted temperature field is a bit complex in fact it is also characterized by a cold drop over the western Europe that is likely to perturbate the $500hPa$ temperatures also in the western Mediterranean. Over the northeastern Italy and the Slovenia the overall lapse rate is estimated to be $33°C - (-10°C) \approx 43°C$. The resulting overall moist adiabatic lapse rate produced by the product of γ_m by the $500hPa$ geopotential height, $5800m - 5850m$, results $\approx 41°C$. This suggests the forecaster that the environment is particularly prone to the instability over the western Europe, but not so much on the northeastern Italian corner. Anyway the forecaster should consider that the real lapse rate could increase suddendly in the upper troposphere because of the advection of colder air due to the south-eastward motion of the cold drop and she/he has to take in account that the lapse rate increase aloft is likely to occur when the maximum temperatures at the ground are reached, because of the insolation. The resulting forecast is that widespread deep atmospheric convection is expected on western Europe while isolated convective events, possibly of strong intensity, are likely in the second half of the day also in northeastern Italy and Slovenia.

Figure 6. METEOSAT visible channel image at 12 UTC of June 29, 2005. There is deep convection over central-western Europe and it is starting on the Italian Northeastern corner. Local severe weather events occurred in the Friuli plain in the afternoon and in Slovenia during the evening (copyright 2005 EUMETSAT)

will be considered in the next paragraphs because generally the information produced by nowadays numerical models outputs have not the required quality for a useful assessment of the mesoscale lower level thermodynamic properties 72 or 36 hours ahead.

Just an example for mid latitudes. If the temperature difference between the ground and the $500hPa$ isobaric level ($\approx 5.5\ km$) is larger than $5.5km \cdot 7Kkm^{-1} = 38.5K$, lets round it at 40 K ($40°C$), then the forecaster is aware that instability is likely to occur.

The weak point of this simple approach is that local instability is very often generated by mesoscale processes especially at the lower tropospheric levels. In this case the forecaster keeps the information of the numerical models outputs for the higher levels, i.e. 500 hPa, and uses the ground temperature that he or she considers to be the best estimate for the computation of the laps rate or the overall temperature difference. This is a clear example of the combination of subjective and objective techniques that the operational forecaster uses to identify risks related to deep convection.

The advantage of this instability estimate is that two temperature fields are used only, furthermore they are standard outputs available for all the operational models and in many cases they are available for free on the web. This allows the possibility for simple but easy to do models output comparison. Another important aspect concerns the quality of the 500 hPa temperature field. This field is one of the best forecasted by all the models, also for long lead times. The possibility to integrate objective numerical outputs with forecaster subjective experience is an advantage too.

Further developments of this technique can be seen with respect to two approaches:

a) The first is to explore better the model outputs with respect to the environmental instability. This is done by means of derived variables like potential temperature and using more that two isobaric levels. Instability indexes are very often available for many model outputs especially LAMs. If not available directly in the model outputs, the forecaster can easily compute them if the model outputs are available in digital form, that is he or she can write simple computer programs to extract the information needed for the computation and visualization of the simplest indexes. The GrADS (http://grads.iges.org/grads) free software is a powerful tool for that purposes.

b) The second one concerns the statistical stratification of the temperature gradients with respect to the climatology dividing them into two complementary classes referring to deep convection occurrence an no deep convection occurrence respectively. The forecasters can statistically identify the suitable threshold for the 500 hPa - ground temperature difference that gives her/him the reference value against which to compare future cases. This will help in case of complex terrain situations. A statistical approach is useful in case of no direct or indirect forecast experience over the interested domain. The climatological values will give more awareness to the forecaster on how to consider situations that are not clearly defined on the base of the above reviewed principles.

Under the light of the temperature gradient considerations, of course advections of cold and warm air are important phenomena. In fact if the forecaster can use more than one vertical level then further capabilities are allowed. In case of cold air advection that are stronger at the lower tropospheric levels, lets say 700 hPa, than that at 500hPa, Then the forecaster can argue that the vertical temperature gradient could be unstable at the lower tropospheric levels, but stable aloft. In this situation deep convection is not likely to occur.

Also the time correlation between the advections at the different levels are important, making unstable situations in which the cold air is coming before at the higher levels than that at lower ones and on the contrary a cold air advection before at the lower levels may reduce or cancel the instability. In some cases the study of the differences in the advection among the available isobaric levels are important elements for the instability evaluation quite in advance.

LAMs give a finer spatial description of the thermal fields and the orography of the interested domain is better represented than in global models, so the forecaster has more grid point on which to work for the instability evaluation and he can attempt to evaluate instability differences over the domain up to scales that are comparable to those typical of the mesoscale. Several LAMs give indications of instability at the mesoscale, in many cases specifying the type of the probable convective phenomena. See figure 7.

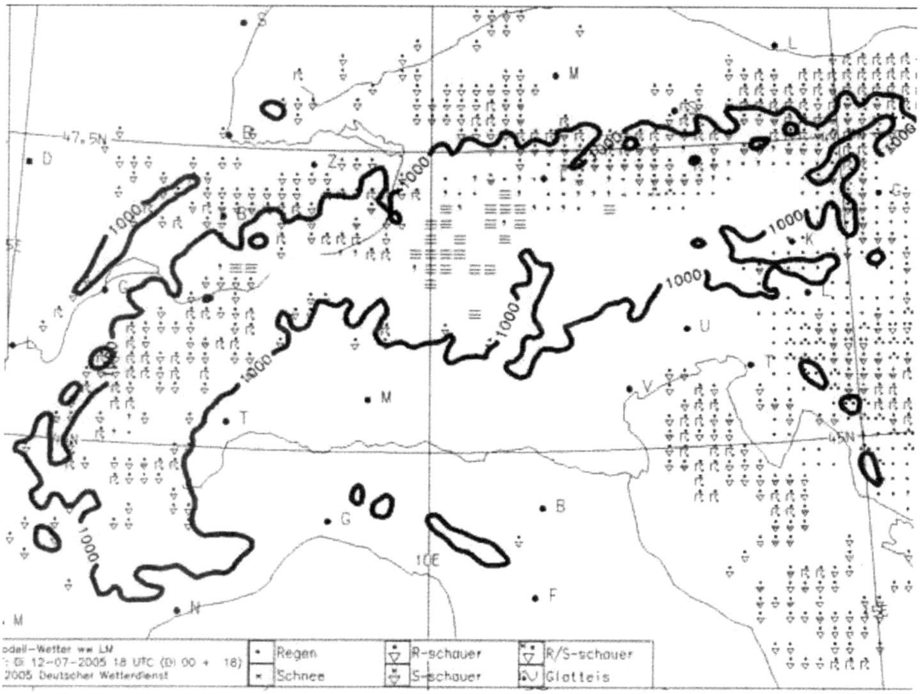

Figure 7. Limited Area Model outputs forecasting the weather type. Here is presented an example from DWD Lokall Model (copyright 2005 DWD).

3.2 Forecasting the trigger

According to the definition of instability, the air parcel will experience the buoyancy force if it will be displaced from its original position. Every dynamical process that lifts the parcel before it starts to experience the buoyancy force is called the trigger of

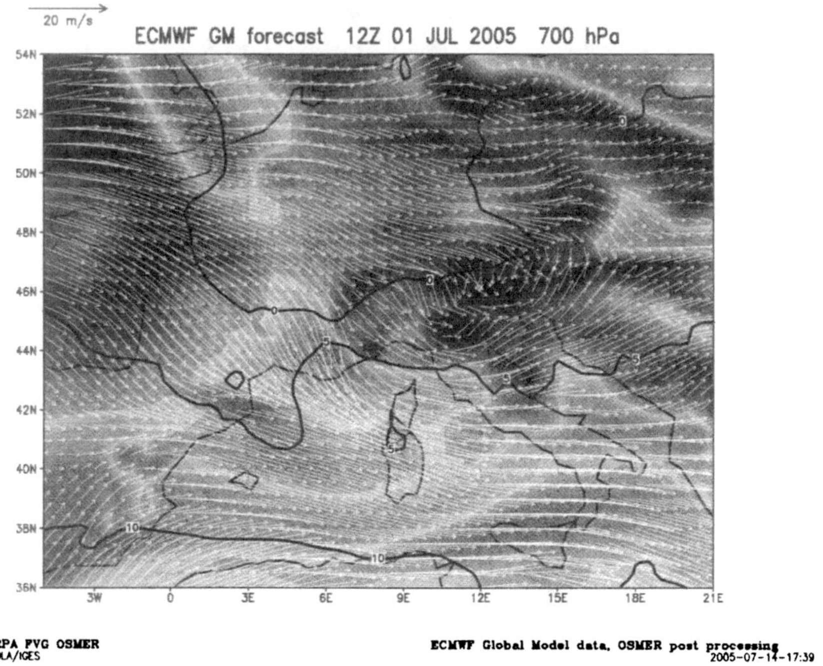

Figure 8. Example of Global numerical model output at 700 hPa simulating the cold frontal passage over the Alpine region. Relative humidity is shaded; darker areas are interested by high relative humidity values, while clearer zones are the dryer ones. Isotherms are the solid black lines. Arrows report the horizontal component of the wind in every model grid point.

the convection. It is the preliminary action that triggers off the spontaneous convective motion.

Triggers are of several types and in many cases they are related to mesoscale processes that are poorly represented in the numerical models. Here, the focus of our attention is on the triggers that could be recognized in operational numerical model outputs and more detailed discussion on triggers that could be identified by real data and experience are left to further paragraphs of these lectures.

One of the most important and common trigger for the convection is the lifting of the lower layers of the troposphere due to frontal action. In fact when a cold front is moving along its track, in the frontal area, the cold air is rapidly replacing the warm air that is lifted. It is well known that the cold air is denser that the warm one so the air replacement along the cold front starts from the bottom of the troposphere. For these reasons it is very important to identify the cold frontal zones. Global models and limited area ones plots automatically the frontal lines at the ground by means of algorithms applied to the basic models outputs. Of course the forecaster can do the frontal identification by

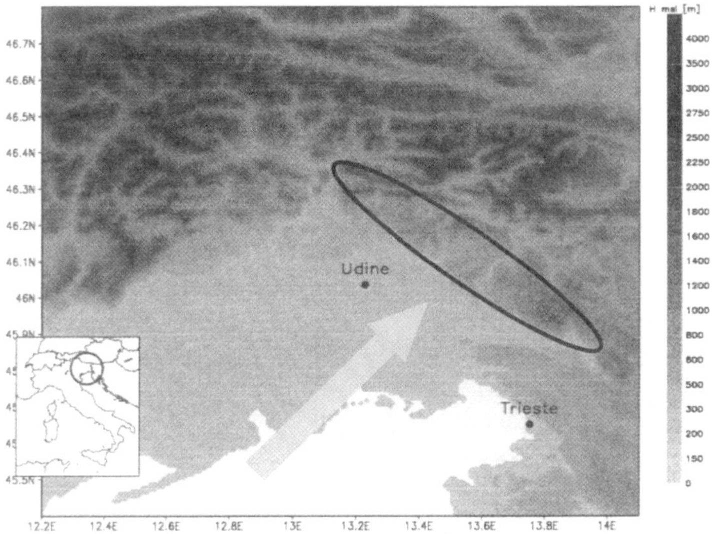

Figure 9. Schematic picture of the synoptic forcing because of the orography. When synoptic winds blow towards a mountainous barrier, it is expected that low tropospheric layers are lifted in the upwind area close to the barrier. In this picture the arrow represents the low level synoptic winds (ground, 925 hPa, 850 hPa), while the ellipse marks the area that the forecatser subjectively considers prone to the convection trigger.

himself, subjectively, using the thermal fields at the different levels or better by means of cross sections of potential temperature.

Over complex orographic areas, like the Alpine region, it is not easy to identify the frontal area at the ground, so one of the possible alternative solutions is to identify the front at an upper level where the complexity of the temperature and wind fields is reduced. An example is the temperature, moisture and wind fields forecasted at 700 hPa, see figure 8. At that level the interaction of the front with the alpine orography is still important but from the forecasting point of view the evolution of the frontal zone is clearer and easier to follow, step by step, in the time sequence of model outputs.

Another trigger that could be forecasted well in advance is the due to the horizontal wind convergence at the low levels, lets say at 925 hPa or 850 hPa, and at the ground of course. In fact, because of the continuity equation, that for our purposes can be well approximated assuming the no divergence of the wind field (Holton, 1972);

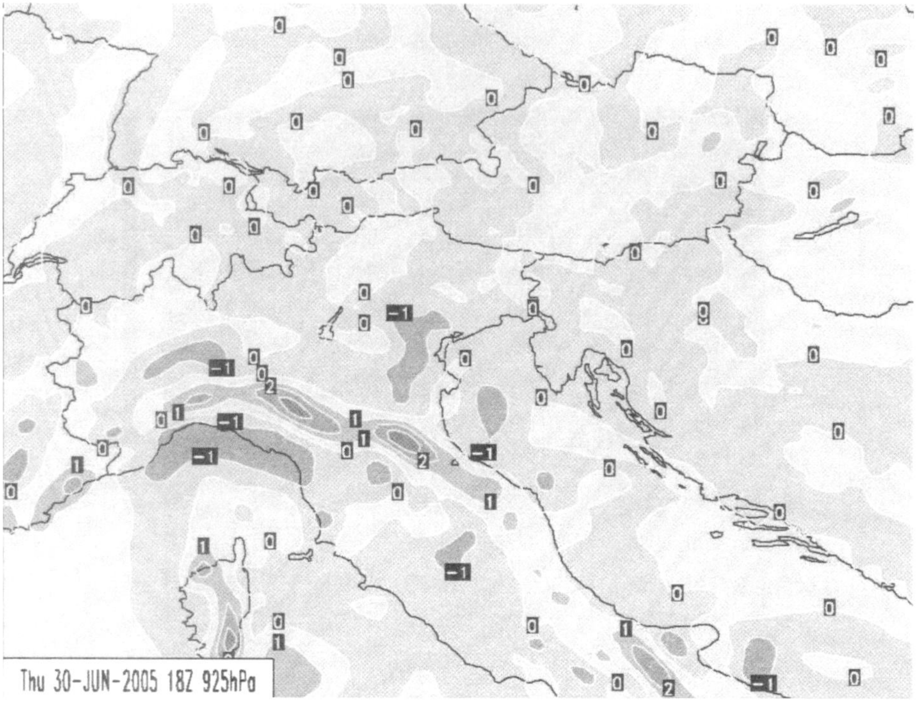

Figure 10. Example of vertical velocities at 925 hPa forecasted by Aladin LAM. Negative values represent the downward component, while positive ones are for lifting air. Isopleths are plotted every 1 $Pa\,s^{-1} \simeq 0.1ms^{-1}$. It is worth to note that the vertical motions are significantly related to the orography of the Appenines. At this level the horizontal wind, here not reported, is mainly blowing from southwest.

$$\nabla \cdot \mathbf{v} = \frac{\partial u}{\partial x} + \frac{\partial v}{\partial y} + \frac{\partial w}{\partial z} = 0 \tag{3.9}$$

here expressed in cartesian coordinates, where $\mathbf{v}$ is the wind vector $\mathbf{v} = (u, v, w)$. If there is horizontal convergence at the lower tropospheric levels, then the flux conservation needs a divergence of the wind along the vertical axis. But, since the air mass at the ground is bounded by the earth surface, that is no air is entering into the soil or the sea, consequently air has to be lifted.

$$\frac{\partial u}{\partial x} + \frac{\partial v}{\partial y} < 0 \Rightarrow \frac{\partial w}{\partial z} > 0 \tag{3.10}$$

This identity shows that the vertical velocity of the air is increasing with height, that is a net transport of mass upwards takes place as a consequence of the horizontal wind convergence. This effect becomes very important when there are interactions between synoptic low level winds and the orography, as in the case of the southern part of the

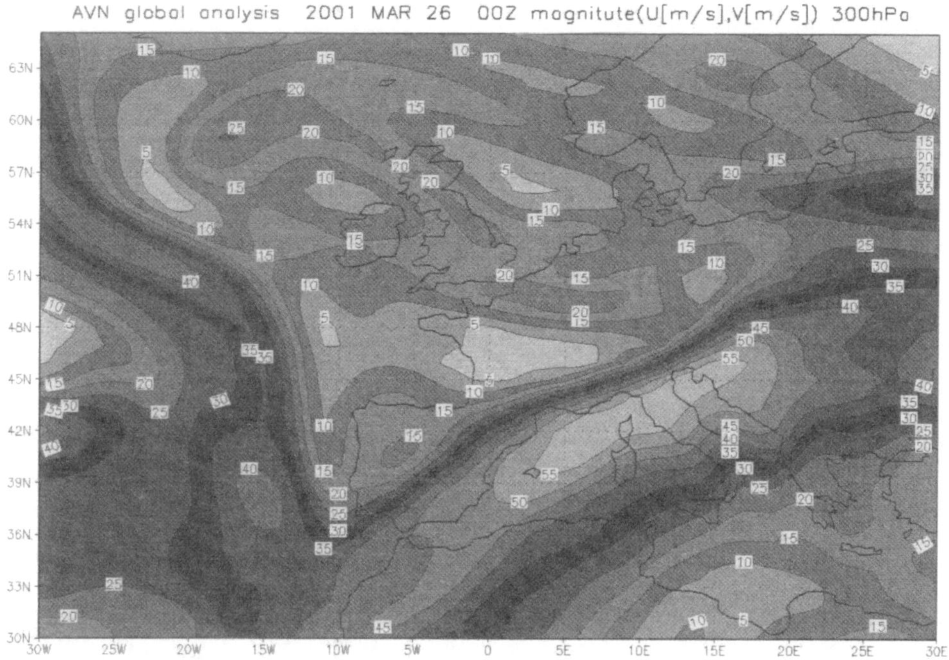

Figure 11. Example of jet over the Northern Italy and western Mediterranean. Dark shaded areas cover large horizontal wind shear zones. Labels report wind speed in ms^{-1}.The plot has been produced by menas of AVN analysis at 300 hPa and maximum speeds exeed $55\,ms^{-1}$. Note the strong horizontal wind shear over the alpine region. Numerical models simulate these field with good quality quite in advance. This environment has hosted a tornado on the Friuli plain.

alpine ridge. When a cold Atlantic front is approaching the Alps, then the southern winds blowing in the warm sector of the frontal area push air masses towards the mountainous barrier generating convergence. Along the foot of the barrier it is expected that part of the horizontal wind momentum will be converted in the vertical one. See Smith (1979), Smith (1989), Baines (1995) for further readings.

Of course this is a qualitatively description of the trigger, anyway in some LAMs, those having a realistic representation of the orography, it is possible to identify the lifting action by means of the vertical velocity fields.

In many cases LAMs outputs allow to estimate the intensity of the sea and mountain breezes. Also in this cases it is possible to evaluate qualitatively the triggering action of the low level winds from numerical models outputs 48 hours or 24 hours ahead. Of course it depends on the model quality. Furthermore, forecasters experience suggests that there are cases of triggers that could take place at the mid or upper levels, gravity waves for

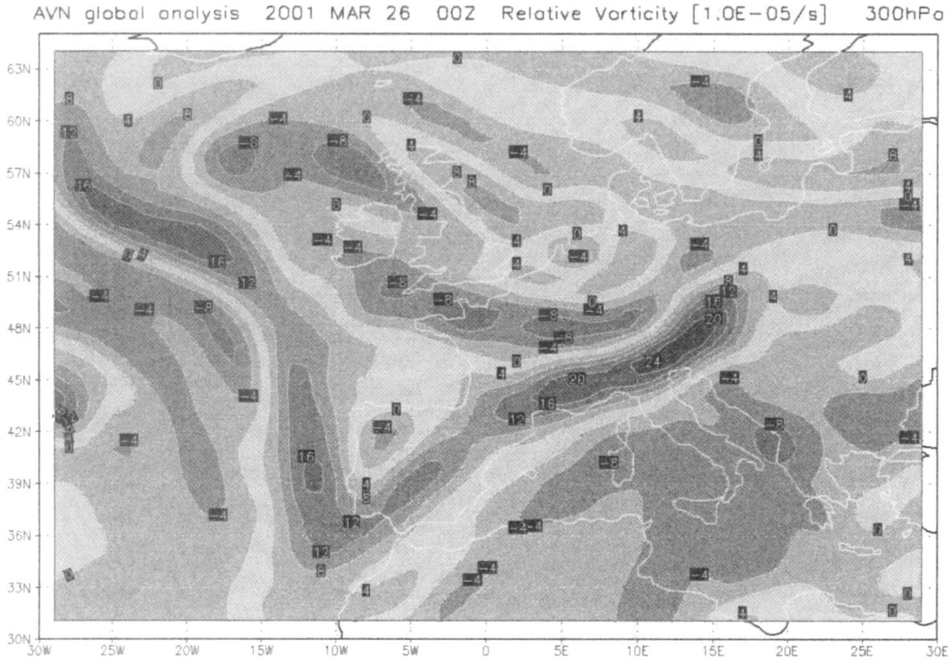

Figure 12. The same as for figure 11, but the vertical component of the relative vorticity is shaded. Dark colors mark the areas with larger positive vorticity values ($> 2.0\,10^{-4} s^{-1}$). Note that the voticity maximum is associated to the strong horizontal wind shear area.

example, produced by the interaction of the upper level winds with the orography. These triggers have already some qualitatively conceptual model proposed but operationally there are just rough ideas on when they could occur.

The trigger of the convection as seen from the numerical model outputs is a promising topic for the development of downscaling tools that could help the forecaster in the trigger identification.

3.3 Forecasting the boundary conditions by means of models outputs only

The boundary conditions that transform deep convection in threatening severe weather episodes are difficult to access 72, 48 hours ahead and in some case 24 hours too, on the base of numerical models outputs only. Many of the important factors that are considered responsible for generating favorable conditions for the severe weather are related to the lower tropospheric layers that are not well simulated in their details by models. For more detail see Giaiotti et al. (2001), Giaiotti et al. (2003), Monteverdi et al. (2003),

Rasmussen et al. (2000), Rasmussen (2003), Doswell et al. (1996), Markowski and Straka (2000), Weisman and Rotunno (2000), Klemp (1987). Anyway there is an important boundary condition, that has been observed many times in combination with severe weather, Bechini et al. (2001), Bertato et al. (2003), and that is under the possibilities of a reliable numerical forecast quite in advance. This condition is a wind jet at the upper troposphere, lets say from 400 hPa to 200 hPa. The basic idea is that the jet at the top of the troposphere generates divergence areas inside which the outflow of the updraft is enhanced.

This is one of the possible exlanations for strong intensity updrafts lasting for a long time. Jets are well forecasted in advance by models and the typical values observed in combination with severe weather events are ranging from 40 ms^{-1} to 60 ms^{-1}.

Bibliography

P. G. Baines. *Topographic effects in stratified flows.* Cambridge University Press, 1995.

G. K. Batchelor. *An Introduction to Fluid Dynamics.* Cambridge University Press, 1994.

R. Bechini, D. B. Giaiotti, A. Manzato, F. Stel, and S. Micheletti. The june 4th 1999 severe weather episode in san quirino, italy: a tornado event? *Atmospheric Research*, 56:213–232, 2001.

M. Bertato, D. B. Giaiotti, A. Manzato, and F. Stel. An interesting case of tornado in friuli-northeastern italy. *Atmospheric Research*, 67–68:3–21, 2003.

C. A. Doswell, H. A. Brooks, and R. A. Maddox. Flash flood forecasting: an ingredients-based methodology. *Wea. and Forecasting*, 11:560–581, 1996.

J. A. Dutton. *Dynamics of the Atmosphere motion.* Dover Publication Inc., 1995.

D. Giaiotti, E. Gianesini, and F. Stel. Heuristic considerations pertaining to hailstone size distributions in the plain of friuli venezia giulia. *Atmospheric Research*, 57:269–288, 2001.

D. B. Giaiotti, S. Nordio, and F. Stel. The climatology of hail in the plain of friuli venezia giulia. *Atmospheric Research*, 67-68:247–259, 2003.

A. E. Gill. *Atmosphere - Ocean Dynamics.* Academic Press, 1982.

J. R. Holton. *An Introduction to Dynamic Meteorology.* Academic Press, 3-rd edition, 1972.

K. Hutter and K. Jóhnk. *Continuum Methods of Physical Modeling.* Springer-Verlag, 2004.

J. B. Klemp. Dynamics of tornadic thunderstorms. In *Ann. Rev. Fluid Mech. Volume 19*, pages 369–402. Annual Reviews Inc., 1987.

P. M. Markowski and J. M. Straka. Some observation of rotationg updrafts in low-buoyancy, high sheared environments. *Monthly Weather Review*, 128:449–461, 2000.

J. P. Monteverdi, C. A. Doswell, and G. S. Lipari. Shear parameter thresholds for forecasting tornadic thunderstorms in northern and central california. *Wea. and Forecasting*, 18:357–370, 2003.

E. N. Rasmussen. Refined supercell and tornado forecast parameters. *Wea. Forecastiong*, 18:530–535, 2003.

E. N. Rasmussen, S. Richardson, J. M. Straka, P. M. Markowski, and D. O. Blanchard. The association of significant tormadoes with a baroclinic boundary on 2 june 1995. *Monthly Weather Review*, 128:174–191, 2000.

R. D. Smith. The influence of the mountains on the atmosphere. In R. Dmowska and B. Saltzman, editors, *Advances in Geophysics Volume 21*. Academic Press, 1979.

R. D. Smith. Hydrostatic airflow over mountains. In R. Dmowska and B. Saltzman, editors, *Advances in Geophysics Volume 31*. Academic Press, 1989.

M. L. Weisman and R. Rotunno. The use of vertical wind shear versus helicity in interpreting supercell dynamics. *J., Atmos. Sci.*, 57:1452–1472, 2000.

General considerations on the operational forecasts of severe convective events: from short range to nowcasting

Dario B. Giaiotti and Fulvio Stel
ARPA FVG OSMER, Regional Meteorological Observatory, Visco (UD), ITALY

Abstract
In this lecture we deal with the general aspects of the convective severe weather forecast in the short range, that is from 24^{h} ahead the occurrence of the event down to the threatening phenomena onset. When the forecast is very close to the event it relates to, lets say a few hours before, or it aims to track the event evolution in real time, the word nowcasting is used instead of forecast. The main elements relevant to issue a local severe weather forecast in the short range are here described, furthermore the information available for that purpose is analyzed with special attention to the limits and the constrains imposed by the operational forecasting activity. Some examples of environments prone to the severe weather onset and evolution are presented and discussed.

1 Nowcasts and forecasts

All the attempts to predict the future state of atmosphere, and the related weather phenomena, are defined as weather forecasts. From the historical point of view, early on, operational forecasts have focused their attention on the evolution of the weather for the next few days. This narrow time window was limited upward by the chaotic behavior of the atmosphere and the very limited performance of the numerical simulations. On the lower extreme of that time range, slowness in the collection and the transfer of meteorological data have limited for many years the extension of the forecasts interest to lead times below the 24^{h}. Anyway, in the past two decades, very important improvements in the numerical models and the data assimilation systems have allowed satisfactory weather forecasts extending for several days, at least for the synoptic atmospheric features. Furthermore, the development of networks of automatic weather stations at the mesoscale producing high frequency and high resolution measurements, and the remote sensing imagery availability, almost in real time, support the possibility to forecast the evolution of the weather during the last hours, just before its real manifestation. So the earliest narrow window has been largely extended and now it is practically bounded upward only. Since the techniques related to the weather forecast dealing with phenomena that will occur in the next hours are mainly based on the observations, their interpretation and the conceptual models developed for the extrapolation of the future weather evolution on the base of those observations, while forecasts related to longer times greatly benefits of numerical simulation outputs, it is quite straightforward to distinguish these two different approaches. For the purposes of this course, the nowcast is defined as the forecast that considers the possibility for a convective severe episode occurrence, at most in the next 24^h, on the base of observational data and measurements only, or at most the first output of the numerical models. Actually, operational numerical models produce simulations very close to the reality for short lead times, that is $+6^h$ or $+12^h$ even at the mesoscale, so very often these model outputs play an active role in the nowcasting process.

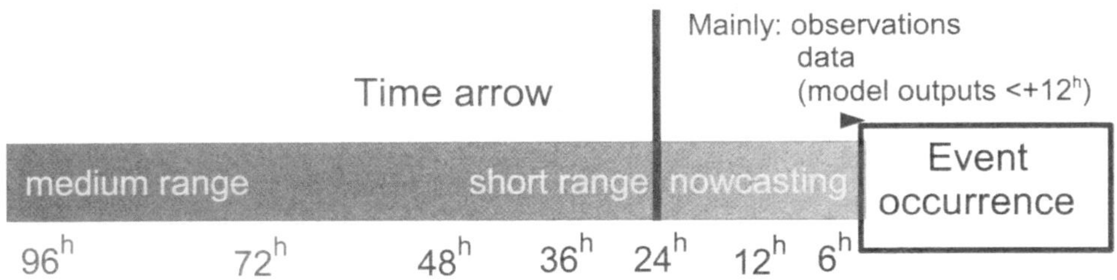

Figure 1. Nowcasting is defined as the weather forecast process that aims to forecast the future state of the atmosphere, and the related weather phenomena, at most 24 hours ahead, but usually only a few. This is because the relevant and reliable information for the evaluation of a severe convective episode, at mid latitudes, are mainly available almost from real data. So, nowcasting is mainly based on observations, measurements, their interpretation and extrapolation. Sometimes numerical models outputs are included in the process but for short lead time only ($<+12^h$).

2 Necessary conditions for convective severe weather occurrence

The onset and the evolution of convective severe weather requires the existence of all the three basic ingredients: a) **tropospheric instability**; b) **convection trigger**; c) **boundary conditions of severe weather occurrence**. As a consequence the forecaster should be able to recognize the existence or the absence of those necessary conditions. The approach to the instability evaluation is generally the easiest, in fact the vertical profile of the troposphere is well monitored at the synoptic scale. As a consequence, over areas without complex orography, the information produced by classical operational in situ measurements, namely radiosoundings, give a pivotal and very useful support for the instability evaluation, even at the mesoscale. Of course mesoscale and microscale processes strongly characterize the lower parts of the troposphere, let's say some hundred of meters closest to the Earth surface. So, in general, a synoptic vertical profile is very useful for the mesoscale description of the higher levels of the troposphere, while it can become poorly representative of the lower troposphere when it refers to a location significantly far, according to the typical mesoscale horizontal distances. This situation is particularly evident in complex orography areas or special geographical locations where the meteorological fields are strongly conditioned by the ground boundary conditions. Information concerning the lower troposphere are usually available from automatic weather station mesonetworks or remote sensing measurements. The hard task of the forecaster is that of bridging the time-space discontinuity existing between the information available for the upper and lower tropospheric levels at mesoscale and microscale resolutions for the purposes of the instability description.

Concerning the trigger for the convection onset and the boundary conditions suitable for the severe weather occurrence, those features are mainly tied at mesoscale and microscale very dynamical processes, so their identification is strongly dependent on measurements and observations at high resolution in space and time. Furthermore the experience of the forecaster plays an important role in the interpretation of the mesoscale weather situation and the extrapolation of its evolution in the next few hours.

3 Evaluation of the instability

The best tool for the evaluation of tropospheric instability is the radiosounding profile. In situ measurements of the main thermodynamic variables, such as temperature, pressure and relative humidity, plus the horizontal component of the wind vector, allow a complete description of the actual thermodynamic state of the air mass, furthermore vertical stratification of the troposphere can be evaluated quantitatively. The theory of the troposphere stability is well developed and it gives objective and quantitative answers to the question concerning the spontaneous vertical air motions, see Zdunkowski and Bott (2004), Iribarne and Godson (1981) and Bohren and Albrecht (1998). From the historical point of view, the radiosounding data have been analyzed by means of aerological diagrams, that is Cartesian coordinate plots in which two fundamental thermodynamic variables are used as abscissa and ordinate. On the plot, radiosounding measurements are reported together with special isopleths, such as isobars, isotherms, adiabats, etc.. There are many kind of diagrams, each one developed for special purposes that let easier the interpretation and exploration of the air properties (Iribarne and Godson, 1981, in chapter VI). The advantage of the aerological diagrams is the possibility to evaluate global tropospheric properties in a glance. Furthermore some diagrams are built in such a way that important integral or differential thermodynamic functions are easy computed graphically, by means of the comparison of selected areas in the plot. Of course nowadays this feature is not so important as in the past, in fact the numerical computation allows a quantitative and fast evaluation of the required quantities. Because of their useful and easy application in the quantitative evaluation of troposphere properties, the aerological diagrams have become default tools in operational weather forecasts, and in particular in nowcasting techniques.

The most common and worldwide used diagram is the so called skew-T-Log P (Diurić, 1994), see figure 2, in which the logarithm of the pressure is reported in the ordinate axis and the temperature is along the abscissa, but the isotherms are not normal to the isobars, they are tilted, and this gives the name skew-T. In recent years, for purely convective forecasts purposes, a new aerological diagram has been defined by Morgan (1992). Its name is Thetaplot and it presents the advantage that conditionally and convectively unstable tropospheric situations are easily recognized in the plot of the equivalent poten-

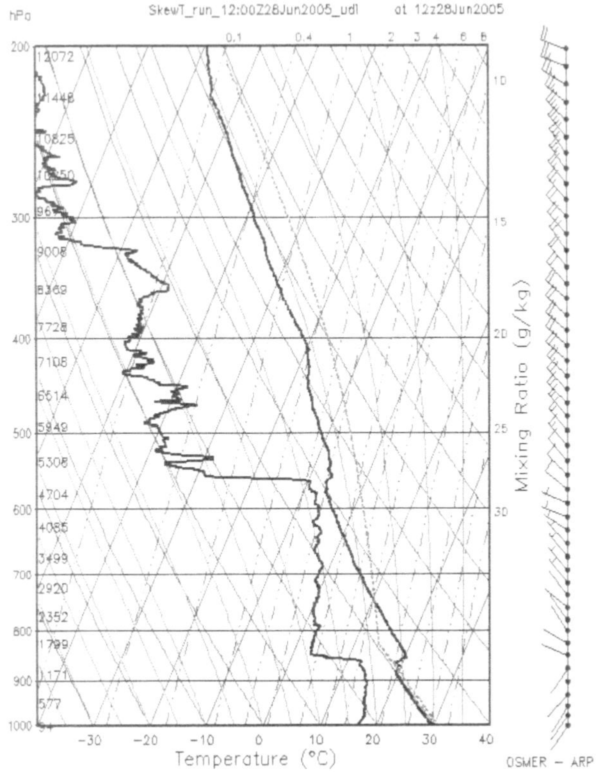

Figure 2. The Skew-T - Log P aerological diagram for an unstable troposphere. The plot refers to the radiosounding data measured by the WMO 16044 station in Campoformido (Udine, ITALY). On the right of the diagram wind barbs report the intensity and the direction of the horizontal component of the wind, by means of which it is easy to recognize levels characterized by particular wind shear patterns. See Diurić (1994) and Iribarne and Godson (1981) for details on the basics of this diagram.

tial temperature which is computed by means of the radiosounding measurements. The potential temperature is plotted against the logarithm of the pressure, see figure 3.

The use of the aerological diagrams for forecasting purposes is based on the assumption that a severe convective event is a mesoscale process that develops in an environment prone to its manifestation and that the radiosounding information will represent the measured environment for at least few hours beyond the time when the measurement is taken. Generally this assumption matches the reality, so when the forecaster looks at the diagram it is likely that, in the next hours, possible mesoscale perturbations of the troposphere will evolve in the environment described by the diagram the forecaster has in hand at that moment. Of course this technique is acceptable for nowcasting purposes, since it is assumed that the tropospheric environment will be only slightly displaced from the initial state, that is described by the radiosounding data. This is a common concep-

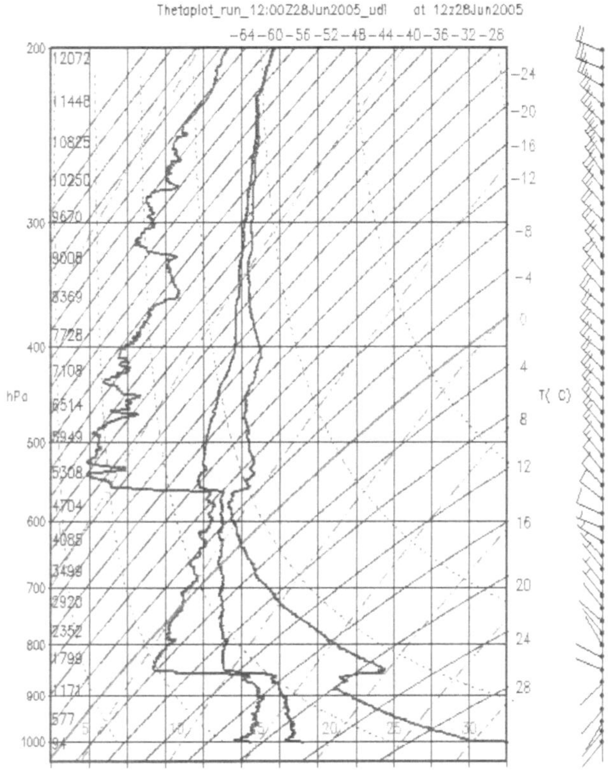

Figure 3. The Thetaplot aerological diagram for the same radiosounding data reported in figure 2, See Morgan (1992) for details on the properties of this diagram.

tualization that permeates the subjective approach to the weather forecast, not only the use of the radiosounding data. The assumption that the evolution of the troposphere is linear in the very short period next to the initial conditions is widely accepted by the forecasters and it has its roots in the observational experience of the synoptic weather systems evolutions. Anyway, the linear evolution conceptual model is working well in a small time interval close to the initial state, where the word small means that the interval is much less than the typical life time of the weather system under forecast. In fact for synoptic systems six hours or half a day can be considered suitable upper limits for the validity of the linear model, while for mesoscale processes one or two hours are sometimes too much.

Far from the initial conditions, the chaotic behavior of the atmosphere is evident also from the numerical simulations. In figures 4 and 5 are reported the time series of the 700 hPa temperature and the rainfall forecast respectively of the ECMWF Ensemble Prediction System for the 13E, 46N grid point. Each of the 51 time series lines reports the forecast for the next ten days, generated by the same model, but with initial conditions slightly different. It is easy to see that for the temperature field (figure 4) within the

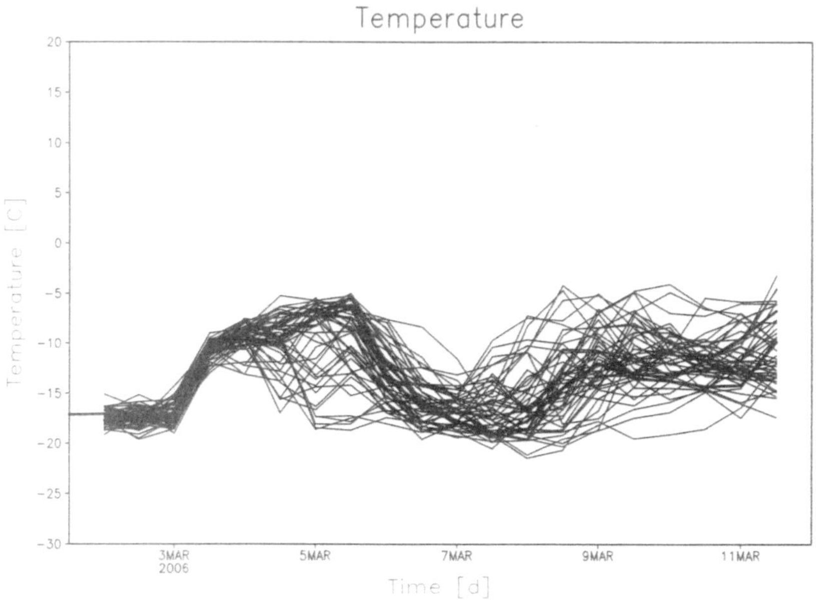

Figure 4. Time series of the temperature forecast at 700 hPa by the ECMWF Ensemble Prediction System. Forecasts are based on the initial state of the atmosphere, that is the analysis, of the 00UTC March 01, 2006 ad each line represent the evolution of the temperature at the 13E, 46N grid point as simulated by the ECMWF operational numerical model but with initial condition slightly displaced from the analysis. For more details on how ECMWF Ensemble Prediction System works see Molteni et al. (1996).

three days closer to the initial state, all the simulations forecast a temperature with a dispersion at most of three degrees. Later than three days the dispersion of the forecasts is wider and they reach a range of about $15℃$. The non linear effect is more evident when derived fields are analyzed, such as the precipitation. In figure 5 it is clear that small displacements in the initial conditions produce rainfall forecasts at +6 days ranging from no precipitation at all to more than 50 mm 6 hours accumulation. For more details on the ECMWF Ensemble Prediction System see Molteni et al. (1996), Chessa and Lalaurette (2001) and Toth et al. (2001). For the non linear problems in the atmosphere, both from the theoretical and numerical simulations point of view see Gilmour et al. (2001), Lorenz (1985) and Lorenz (1963).

By means of radiosounding data, the forecaster can evaluate the overall instability of the troposphere looking at some worldwide known and well defined variables and indexes.

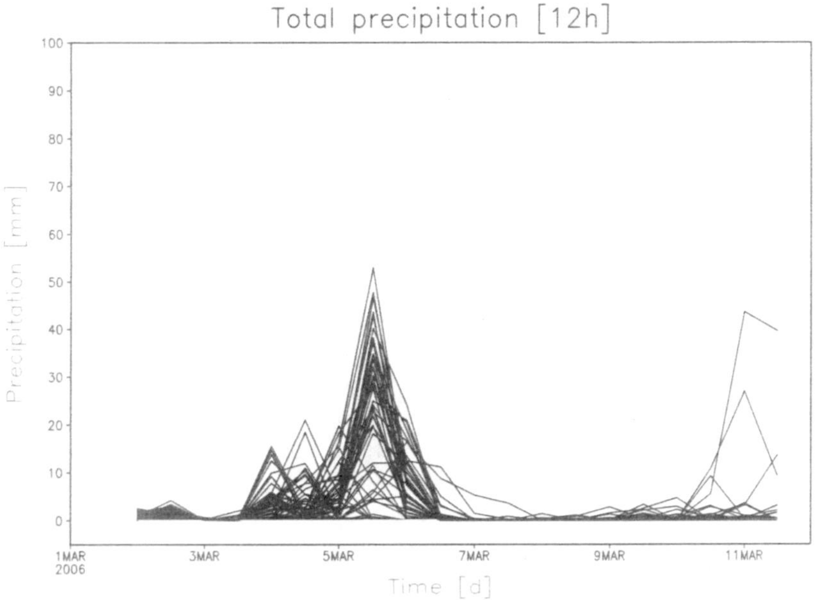

Figure 5. Time series of the precipitation forecast at the 13E, 46N grid point as simulated by the ECMWF Ensemble Prediction System. Forecasts. The initial state of the atmosphere and the model features are the same as for figure 4.

The classical instability variable is the CAPE, Convective Available Potential Energy, see Diurić (1994), that is used in combination with the CIN, Convective INhibition, see Diurić (1994). CAPE is the amount of buoyant energy available to an unitary mass air parcel for its upward acceleration. The higher the CAPE value, the more energy available for a thunderstorm growth. CAPE is especially important when air parcels are able to reach the LFC, Level of Free Convection, that is the level at which a lifted air parcel becomes buoyant. To reach the LFC, generally a parcel have to be lifted because in the lower tropospheric stable environments the parcel is not buoyant, so it has to do work otherwise it sinks back down again. This amount of work is the CIN. In summary, the forecaster has three important quantitative elements to evaluate the troposphere instability that possibility will lead to deep convection occurrence: CAPE telling him how much energy will have an air mass to accelerate upward when it becomes buoyant; LFC indicating to which height the air mass has to be lifted before to become buoyant and be allowed to benefit of CAPE; CIN reporting how much work has to be done by an external agent to lift the air mass from the bottom troposphere to the level where it starts spontaneously to rise. So far there is not a golden rule, involving CAPE, CIN and LFC, that can be applied to discriminate between convective severe weather and normal deep atmospheric convection. Several studies (Nordio et al. (2003), Thompson et al. (2003), Markowski and Straka (2000), Weisman and Klemp (1982) and Lilly (1979)) have been carried out to evaluate thresholds for each of the above variables or their combinations with the aim to define robust operational tools for the nowcasting of severe convective storms or classes of them, such as tornadic supercells. Those studies have pointed out that all those three variables are important, but unfortunately they have not defined a universal and practical quantitative scheme that forecast severe storms with a skill higher than the subjective nowcast. Furthermore experience suggests that severe weather can occur even with very low CAPE values, that is with a very low availability of energy favorable to the air buoyancy (Bertato et al., 2003).

Subjectively, the forecaster looks at the CAPE and verifies that it is not zero, but if it is greater than a few hundreds of $J\,kg^{-1}$ it recognizes that there is energy available for buoyancy, possibly deep atmospheric convection. Furthermore taking in account the LFC the possibility to have some suitable lifting mechanism is evaluated by means of experience and the knowledge of the mesoscale and microscale phenomena, such as breezes, local or synoptic winds and their interaction with the orography that could generate the needed upward forcing to let the air parcels to reach the LFC. The CIN contributes to the evaluation of the forcing strength and duration required to act on the air parcel to do the needed work.

Besides the above mentioned basic elements, many other indexes have been developed with the aim to get predictors for severe weather occurrence due to convection. Almost all of them are the result of considerations on the temperature and moisture vertical gradients, in some cases they include information on horizontal wind speed or vertical wind shear. In many cases they have not a clear physical meaning because they are the result of a statistical regression, where the regression is made over almost all the meteorological variables at all the levels available from a radiosounding. Among then the most popular are the K index, the Totals index, the SWEAT index and Lifted index. For more details on the definition of the above mentioned indexes see Diurić (1994). Some indexes have been obtained and tuned for special areas of the world, for example the Friuli Venezia Region (Italy), where the EV index and the COIN index are currently used in daily operational weather forecasting activity, see Giaiotti and Stel (2001) and Epifani and Vento (1978).

Nowadays, all the instability indexes are easily computed numerically from the radiosounding data and they are reported in combination with the thermodynamic diagrams, see figure 6 as an example. This graphical and numerical combination of information is the most important tool for subjective nowcast of tropospheric instability.

Since current operational numerical models outputs have an high vertical resolution, that is many vertical levels are available, it is possible to produce virtual radiosoundings. Virtual radiosoundings are vertical profiles of the troposphere over a selected geographical location, obtained from the numerical models outputs. They are also called pseudoTEMP, because they are like TEMP measurements reports, but they are not real measures. So the same forecasting technique applied to the real radiosounding is used on the virtual radiosoundings, allowing the finer nowcasting evaluation of instability to

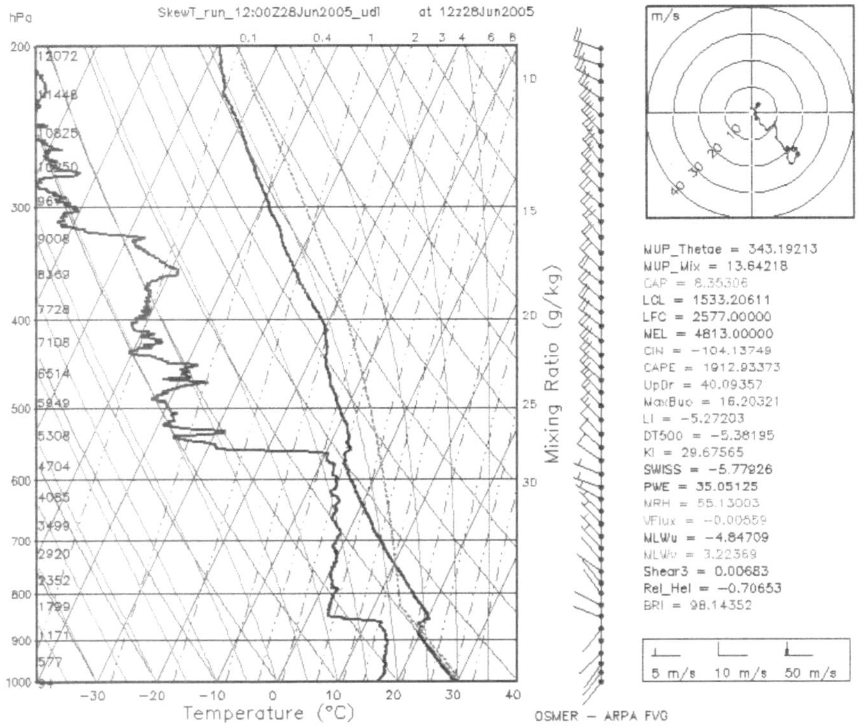

Figure 6. Example of the combination of thermodynamic diagram, Skew-T - Log P in this case, and the list of the instability and severe weather indexes computed from the radiosounding data. The indexes values are reported on the right of the plot and they have different colors according to their position in comparison with the empirical distribution of the corresponding index value, that is not common values are highlighted. On the upper right corner the hodograph completes the plot giving information on the wind shear.

be extended further beyond the implicit temporal limit imposed by the assumptions on the linear evolution of the mesoscale weather systems and the availability of the TEMP reports. This is very helpful for the improvement of the short term forecast quality. Of course, to the limited reliability of the simulated profiles, the uncertainty of the technique has to be added, so in principle the instability forecast based on pseudoTEMP analysis has less skill (Wilks, 1995) than instability nowcast based on real sounding data. By means of pseudoTEMP data it is possible to produce the sounding analysis that is done numerically on real sounding, then aerological diagrams and index computation are presented with the same format, see figure 7.

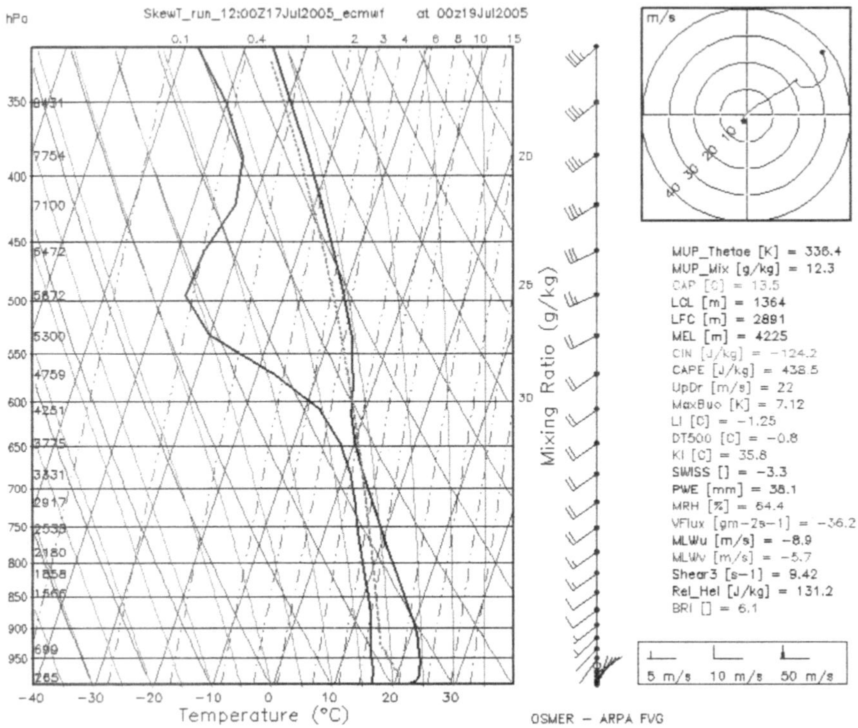

Figure 7. Example of pseudoTEMP analysis. The plot and the indexes have the same format as for the figure 6, but it refers to the 19 July 2005 00UTC pseudotemp that was extracted from the ECMWF $+48^h$ forecast for the location 46N, 13E. In this figure, from the less details available in the temperature lines and the smoothness of the vertical wind shear, it is possible to recognize the limits the model has in representing the mesoscale tropospheric features. Compare this figure with figure 6.

4 Evaluation of the convection ignition

The initiation of convection has to be carefully evaluated to nowcast severe weather occurrence. In fact many times environments characterized by an high instability do not produce any convective event, in those cases it is clear that a trigger is lacking. Mainly, the type of the triggering event and it intensity is subjectively evaluated by the forecaster. To do that the knowledge of the orography and the climatology of geographical domain object of the forecasts is pivotal. For this reason it is very important the forecaster can study previous severe weather events by means of mesonetwork data, radar and satellite imagery, lightning reports and all the other sub-synoptic information that could help him in identifying the causes of the convection ignition. Of course in some cases the convection is stimulated by a pure synoptic forcing, such as a cold front passage, but in complex orographic areas the non pure synoptic triggers are dominant. In fact

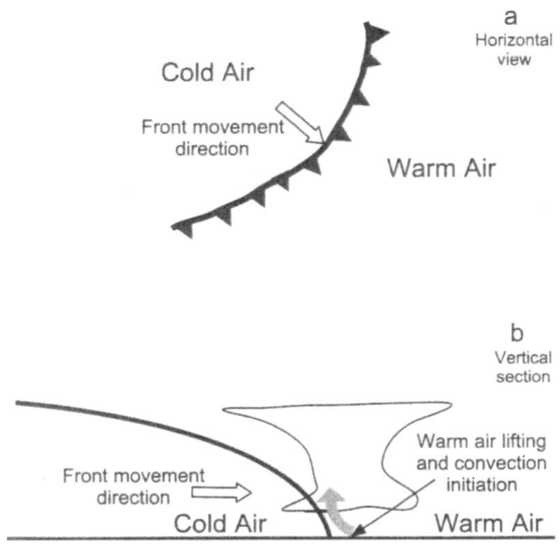

Figure 8. Simple picture showing the area interested by convection ignition due to a cold front passage. The part (a) gives an horizontal view of the frontal area with common meteorological symbols. In part (b) a vertical section of the troposphere is plotted schematically and the frontal area where the incoming cold air lifts the warm air is put in evidence.

also the interaction between synoptic winds and the orography can be considered among the mesoscale phenomena because the resulting interaction is strongly depending on the orography peculiarities that have a typical mesoscale or microscale spatial resolution.

The main synoptic forcing to the convection is the lift of the moist warm air due to the incoming cold air in the frontal zone. Along the frontal area, see figure 8 the cold air is rapidly replacing the warm moist air at the lower levels. This results in the lifting of the air masses close to the atmosphere bottom. In many cases it is easy to recognize the convection development along the frontal area from radar and satellite imagery. The zones characterized by convection ignition are moving with the front. Good numerical forecasts of front development, especially in non complex terrain, can be very useful for the evaluation of the frontal forcing occurrence in time and space. When cold fronts interest areas characterized by complex orography the resulting convection trigger is more difficult to forecast because the frontal area at the surface is not easily recognizable.

The interaction between local winds, that is mesoscale or microscale winds, with the orography is another important cause of the convection ignition. In fact, because of the mass conservation, see equation 4.1, low level atmospheric flows impinging on a orographic obstacle are forced to lift, and to flow around it. If the forced upward motion is sufficient to let the lower tropospheric air masses to reach the LFC, then convection occur. Local winds are usually breezes or winds resulting from the interaction with

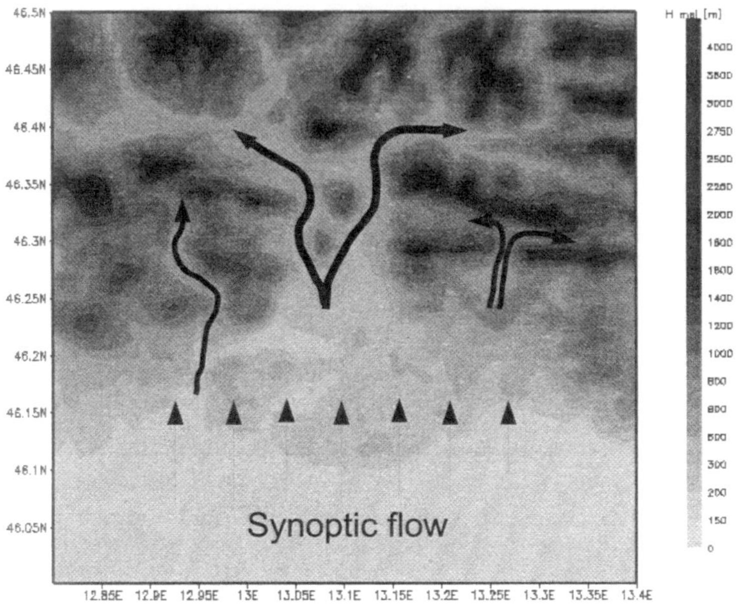

Figure 9. Schematic description of the low levels atmospheric circulation forced in valleys. In this example the deviation of a southern flow impinging the Prealpine area of the Friuli Venezia Giulia region, NE Italy, is qualitatively reported by black arrows. The map reports the orography of the region.

synoptic circulation and orography, such as winds forced to flow in valleys. Figures 9 and 10 show an example.

For an incompressible fluid the continuity equation is:

$$\frac{\partial u}{\partial x} + \frac{\partial v}{\partial y} + \frac{\partial w}{\partial z} = 0 \tag{4.1}$$

with (u, v, w) the three components of the wind vector. Close to the orography the wind cannot penetrate the obstacle so $\frac{\partial u}{\partial x} + \frac{\partial v}{\partial y} < 0$ and, as a consequence, a positive vertical velocity gradient $\frac{\partial w}{\partial y} > 0$ exists.

Deep atmospheric convection can be triggered by strong shallow convection at the ground. This is the case of mountain slopes exposed to the sun radiation. In that case, along the mountain slope warm air is produced continuously and a vigorous shallow convection develops. If the shallow convection take place close to the LFC, then the upward motion of the air masses may be sufficient to initiate the deep convection. Once the deep convection has started then the microscale conditions are largely perturbed and

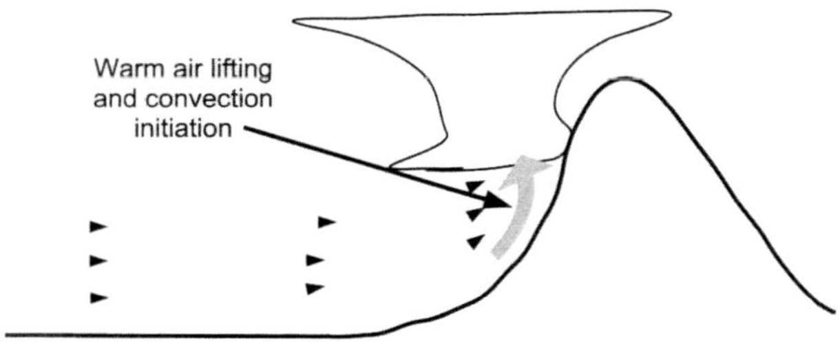

Figure 10. Schematic description of the orographic forced upward motion that is responsible for the convection initiation when low level atmospheric flows impinge on a mountainous barrier.

the feeding of the updraft is extended to areas larger than the mountain slope. See figure 11 as a schematic example.

Last but not least as importance and frequency, is the convection trigger due to the upward forcing of the air close to the ground because of the mesoscale and microscale cold fronts produced by the downdrafts belonging to already existing convective cells. This cause of the convection ignition is very difficult to forecast, and only in special cases, by means of Doppler radar information it is possible to foresee it in the right position and time. This ignition effect is analogous to the mechanism working in the synoptic cold front lifting. In fact cold and dry air reaching the ground by means of the downdraft, cannot penetrate the ground surface, so it is forced to spread around, see equation 4.1. Moving far from the downdraft, that denser air replaces the surrounding one which is vigorously lift.

5 Boundary conditions of severe weather occurrence.

The most difficult ingredient the forecaster has to recognize in nowcasting convective severe weather occurrence is the existence or the development of the proper boundary conditions that transform deep atmospheric convection in severe weather. This is the most important ingredient that discriminate between risky and safe convective weather. The main problem concerning the boundary conditions is that, for many severe weather phenomena, they are not completely understood. Anyway, for some classes, a lot of case studies carried on all over the world and numerical simulations give useful suggestions. In particular for thunderstorms hosting tornadoes, low levels vertical wind shear is considered the principal discriminating factor, see Weisman and Klemp (1982), Klemp (1987), Weisman and Rotunno (2000). In fact the horizontal components of the vorticity vector can be transformed in vertical vorticity because of the interaction between the updraft and the sheared wind environment. Considering the equation for the time variation of the vorticity vertical scalar component ζ:

$$\frac{d\zeta}{dt} = \omega_h \cdot \nabla_h w + \zeta \frac{\partial w}{\partial z} \tag{5.1}$$

where ω_h is the vorticity horizontal component and w is the vertical component of the

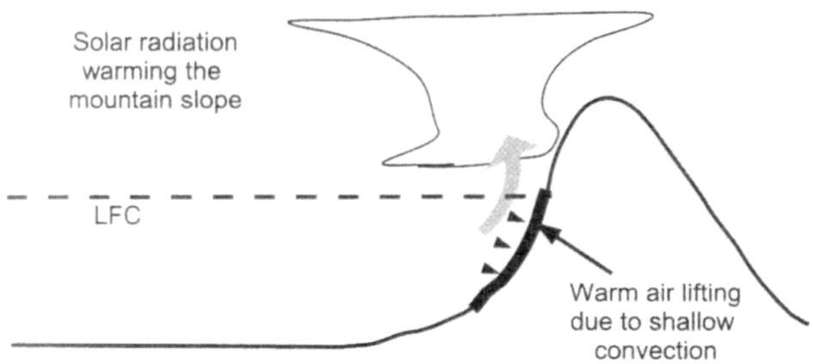

Figure 11. Qualitative example showing the initiation of deep atmospheric convection due to shallow convection occurring on a mountain slope which is exposed to the solar radiation. In the areas of the slope where the exposition to the sun light is close to be normal the radiation flux is large, then the ground temperature reaches high values stimulating a vigorous shallow atmospheric convection. If the air masses involved in the convection are close to the LFC, then deep convection can be initiated.

wind, it is possible to recognize that the time variation of the vertical vorticity is made by tilting of the horizontal vorticity $\omega_h \cdot \nabla_h w$ and by stretching the already existing one $\zeta \frac{\partial w}{\partial z}$. So horizontal vorticity, by means of updraft, contributes in the increase of the updraft rotation, that is giving vertical vorticity to the storm. In equation 5.1 ∇_h represents the horizontal component on the gradient operator and no mixing contribution has been considered; see Klemp (1987) for more details.

The forecaster looks for low level wind shear mainly by means of radiosoundig measurements. An objective evaluation of the shear is made computing wind derived variables such as the classical wind shear in the first 3 km at the bottom troposphere of the Storm Relative Environmental Helicity (SREH), see Diurić (1994). Mesocyclonic thunderstorms have been observed in NE Italian plane with SREH ranging from 50 $J\,kg^{-1}$ to 200 $J\,kg^{-1}$ and in environmental vertical wind shear $> 5.0 \cdot 10^{-3}\ s^{-1}$ and in many cases $> 1.0 \cdot 10^{-2}\ s^{-1}$. Unfortunately, so far, there are no robust statistical thresholds that characterize tornadic or mesocyclonic environments, so the evaluation of the shear influence on developing storms is based on the forecaster experience.

The increase of the horizontal vorticity of the wind field in the lower tropospheric levels can come from local, mesoscale, peculiarities of the temperature and moisture fields. In fact large horizontal gradients, especially in the thermal field, results in marked gradients also in the buoyancy field. If a storm develops across such a gradient, the updraft flowing normal to the gradient experiences an increase of horizontal vorticity that is subsequently transformed in vertical vorticity by means of tilting, as explained above in equation 5.1. Many cases have been studied, for example Rasmussen et al. (2000) and Bertato et al. (2003), and the generation of the baroclinc boundaries where the storms develop and become severe storms are consequence of different mechanism. In some cases they are due to spatial discontinuities in the solar radiation reaching the ground because of the shade produced by the presence of already existing thunderstorms, in other cases they are the result of the cold air outflow from mountain valleys. In all these situations, only mesonetwork measurements with high time resolution can put in evidence the onset of such a boundaries.

The presence of a jet in the wind field in the upper troposphere has been observed in many tornado episodes Bechini et al. (2001), Bertato et al. (2003) but its relation with the severe weather occurrence is less strong than that put in evidence for low level wind shear. The basic idea sustaining the importance of the jet in convective severe weather is that the outflow at the top of the convective cells, that is the summit end of the updraft, is intensified by means of the jet, then updrafts have more probability to stay in stationary conditions. Since the jet at the upper tropospheric levels is generated by synoptic conditions, that boundary condition is easy identifiable by radiosounding profiles or weather analysis and it is a robust feature of the environment that last for several hours.

Nowcast of severe weather connected to large hailstones looks for information in the CAPE of the radiosounding describing the environment where the storm develop. That is large CAPE values imply potentially strong updrafts, so large hailstones can be maintained in their water vapour rich environment for longer time before to fall at the ground. Furthermore, recent studies (Giaiotti and Stel, 2004) have pointed out that high values of mixing ratio (Diurić, 1994), that is water vapor in the air, at the ground level can be considered an important signal for large hail formation.

Concerning flash flood, or local intense rainfalls, for example those events exceeding 50 $mm\, h^{-1}$, there is not any robust practical rule for the nowcasting of the boundary conditions characterizing this kind of risk. In this case, the radar and satellite imagery are the basic elements for the subjective nowcasts.

6 General considerations on severe weather index definition.

The aim to make as soon as objective the convective severe weather forecasts leads naturally to the definition of indexes. Generally, in the simplest form, a severe weather index is an ordered scalar variable that is computed from objective meteorological inputs. Those inputs can come from measurements, numerical simulations or their combination. The basic idea is that the index is monotonically tied with the probability of a severe weather occurrence, so reading the index value means to read the probability to have e severe weather event. In principle the best index would be one having two states only, that is 0 and 1, where 0 means no severe weather and 1 severe weather occurrence. Because of the complexity of the weather it is straightforward to link the index with the probability of the event occurrence. Nevertheless, very often, when the quality of the index is evaluated, comparing its forecasts with the reality, the continuous index domain is projected in a two state variable by means of a suitable threshold. (Wilks, 1995).

There are some simple advices that a index developer should follow to produce an operational severe weather index. First of all it is necessary to evaluate what are the information available to the operational forecast activity. This means that input data, that will be potentially used for the index computation, have to be available in time for an useful operational use of the index. In particular, every time the index is required, that is once or more in a day, inputs should be available and the index computation has to take a negligible time in comparison with future time interval it refers to. In other words, indexes that are very good, but that are available later that severe weather occurrence or that can be computed only if specific data enter into the index computation and those data are produced only in special cases, makes the index operationally useless.

There are two complementary approaches to define the index that some time are used independently. One is mainly deterministic and it is based on physical considerations on the severe weather phenomena the index is voted to and the physical properties of the environment hosting the phenomena. The other considers the severe weather occurrence marked by precursors that are usually present in the severe weather environment; this is mainly a statistical approach that produces an index as the result of the analysis of a large set of past severe weather events. Usually a statistical regression is made. In some cases those complementary approaches can coexist, in fact physical considerations can be used to define the set of variables over which to make the statistical analysis, or the functional form of the index can be defined a priori according to some physical constrain. Neither of the above mentioned approaches can be considered better in principle and both have advantages. Concerning disadvantages, the deterministic approach suffer a lot of the atmosphere complexity and its chaotic evolution. This limits a lot the method for operational purposes. The statistical approach is strongly dependent of the dataset used for the index definition, so many times the resulting index is tuned on the geographical area and the climatic period form which the data have been extracted.

Once the index has been defined, it is necessary to determine its link with the probability of the event occurrence that operationally is considered valuable. In fact, the requirements of the operational weather forecast are not based on meteorological information only, but they take in account other elements, such as the impact of the forecast on the end-users, for example the economic value of the forecasts, the language limits in the forecast communication and other things that are not involved in the index development. Usually this is done looking at the index forecast and the corresponding severe weather occurrence and generally results in a selection of one or more index thresholds that transform the index values in severe weather probability or class of probability of its occurrence. This is accomplished also running the index operationally, but off line for a trial period. That is the forecaster looks at the index, but he does not consider it as an important source for forecast process. During this phase the forecaster reports his subjective impressions on the index and comments on its operational applicability. In some cases he can set the proper set of thresholds subjectively. Generally a proper period of pre-operational trial of the index is always needed.

The last aspect that should characterize an index development is the index verification. When the index is running operationally, it is very important to have developed a verification scheme that tracks the index quality. It is not important that the verification is performed operationally, even if this would be very helpful to follow the index behavior in real time. Anyway even periodical index verification allow the developers to tune the index and to study the response of the forecasters to the index use.

Bibliography

R. Bechini, D. B. Giaiotti, A. Manzato, F. Stel, and S. Micheletti. The june 4th 1999 severe weather episode in san quirino, italy: a tornado event? *Atmospheric Research*, 56:213–232, 2001.

M. Bertato, D. B. Giaiotti, A. Manzato, and F. Stel. An interesting case of tornado in Friuli-northeastern italy. *Atmospheric Research*, 67–68:3–21, 2003.

C. F. Bohren and B. H. Albrecht. *Atmospheric Thermodynamics.* Oxford University Press, 1998.

P. A. Chessa and F. Lalaurette. Verification of the ecmwf ensemble prediction system forecasts: a study of large-scale patterns. *Wea. and Forecasting*, 16:611–619, 2001.

D. Diurić. *Weather Analysis.* Prentice Hall, 1994.

R. Epifani and D. Vento. Studi particolari sulle caratteristiche della grandine nel Friuli-Venezia Giulia. Technical report, ERSA, 1978.

D. Giaiotti and F. Stel. Comparison between subjective and objective thunderstorm forecasts. *Atmospheric Research*, 56:111–126, 2001.

D. B. Giaiotti and F. Stel. Environmental variables affecting the hailstone size distribution. In *Proceedings of the 14th International Conference on Clouds and Precipitation*, pages 1796–1798, 2004.

I. Gilmour, L. A. Smith, and R. Buizza. Linear regime duration: is 24 hours a long time in the synoptic weather forecasting? *J. Atmos. Sci*, 58:3525–3539, 2001.

J. V. Iribarne and W. L. Godson. *Atmospheric Thermodynamics.* D. Reidel Publishing Company, 1981.

J. B. Klemp. Dynamics of tornadic thunderstorms. In *Ann. Rev. Fluid Mech. Volume 19*, pages 369–402. Annual Reviews Inc., 1987.

D. K. Lilly. The dynamical structure and evolution of thunderstorms and aquall lines. In *Ann. Rev. Earth Planet. Sci. Volume 7*, pages 117–161. Annual Reviews Inc., 1979.

E. N. Lorenz. Deterministic nonperiodic flow. *J. Atmos. Sci*, 20:130–141, 1963.

E. N. Lorenz. The growth of errors in prediction. In *Proceedings of the International School on Physics - Enrico Fermi Course LXXXVIII - Turbulence and predictability in Geophysical Fluid Dynamics and Climate Dynamics*, pages 243–265, 1985.

P. M. Markowski and J. M. Straka. Some observation of rotationg updrafts in low-buoyancy, high sheared environments. *Monthly Weather Review*, 128:449–461, 2000.

F. Molteni, R. Buizza, T. N. Palmer, and T. Petroliagis. The ecmwf ensemble prediction system: methodology and validation. *Q. J. R. Meteorol. Soc.*, 122:73–119, 1996.

G. Morgan. Thetaplot, an equivalent potential temperature diagram. *Meteorol. Atmos. Phys.*, 47:259–265, 1992.

S. Nordio, L. Stefanuto, and F. Stel. Preliminary studies on the occurrence of local severe weather events in Friuli Venezia Giulia. *Atmospheric Research*, 67-68:517–522, 2003.

E. N. Rasmussen, S. Richardson, J. M. Straka, P. M. Markowski, and D. O. Blanchard. The association of significant tormadoes with a baroclinic boundary on 2 june 1995. *Monthly Weather Review*, 128:174–191, 2000.

R. L. Thompson, R. Edwards, J. A. Hart, K. L. Elmore, and P. Markowski. Close proximity soundings within supercell environments ontained from rapid update cycle. *Wea. and Forecasting*, 18:1243–1261, 2003.

Z. Toth, Y. Zhu, and T. Marchok. The use of ensembles to identify forecatss with small and large uncertainty. *Wea. and Forecasting*, 16:463–477, 2001.

M. L. Weisman and R. Rotunno. The use of vertical wind shear versus helicity in interpreting supercell dynamics. *J., Atmos. Sci.*, 57:1452–1472, 2000.

N. L. Weisman and J. B. Klemp. The dependence of numerical simulated convective storms on vertical shear and buoyancy. *Monthly Weather Review*, 110:504–520, 1982.

D. S. Wilks. *Statistical methods in atmospheric sciences.* Academic Press., 1995.

W. Zdunkowski and A. Bott. *Thermodynamics of the Atmosphere.* Cambridge University Press, 2004.

Weather Forecast Verification

Fulvio Stel and Dario B. Giaiotti
ARPA-OSMER Regional Meteorological Observatory, Visco, Italy

Abstract These notes are voted to stress the importance as well as the intrinsic difficulty of the weather forecasts verification, giving some hints to solve specific problems and some tools to face various situations. In general weather forecasts cannot be fully wrong but they cannot be neither fully right, this because they are trying to represent a future state of an extremely complex system, which is defined by too many aspects to be fully well described. There is a quite general confidence on the fact that it is at least possible try to quantify the amount of good and bad information that forecasters are trying to give on that future state. Nevertheless it is not possible to define in a unique way this quantification process, then different verifications procedures might give different results even if correctly realized. The standardization of definitions and of procedures is generally still poor and sometimes contraddictory. This fact makes, if possible, even more difficult to deal with the weather forecast verification. Facing the verification of rare weather events, as can be the case for the phenomena related to deep moist convection, extra difficulty arise by the fact that the powerful tool represented by statistics becomes less effective and the intepretation of results becomes in those cases even more tricky. In any case the verification of weather forecasts is an extremely important and structural aspect of the forcasting activity, that cannot be considered complete without it. Moreover the verification of weather forecasts can be an important opportunity to have a different look to the atmospheric aspect toward which we are pointing our attention and, for whom it might interest, to have a different look at the forecasters mind.

1 Introduction

"*Qui vive la pietà quand'è ben morta:*
chi è più scellerato che colui
che al Giudicio Divin passion porta?"
(D. Alighieri - Divina Commedia, Inferno, 20th Song)

The 20th song of the Divina Commedia's Inferno is entirely dedicated to all the "forecasters", that is to people that dedicated their lives in forecasting future events. They are condemned to walk for the whole eternity on a rough terrain with their faces turned back, then without any possibility to look at where they are walking on. In this way, according to the imagination of Dante, they are paying their guilty attempt to look at the future, which belongs only to the will of God ("Giudicio Divin"). At present times, nearly 800 years after Dante, weather forecasters, for some aspects a sort of modern "scellerati" (fools), are condemned to the same pain, that is to walk ahead turning back their faces looking at the already trodden ways. This pain has a name and its name is *forecast verification.* But forecast verification, even if some forecasters live it as a pain, it is definitively something more, in fact it is a fundamental aspect of the forecasting activity at least for two reasons: i) it is the only way in which forecasters can have a feedback on their activity as shown among others in Murphy and Sabin (1986), Vescio and Thompson (2001) and Bosart (2003); ii) it is the only way in which people who are using systematically weather forecasts can obtain the needed informations for their decisions as shown in Le Blancq and Johnson (2003). In general, following the sentence of Brooks and Doswell (1996), we can add that issuing weather forecasts without verifying them it is as managing a factory or a company using only planning budgets without consumptive ones.

Before to start facing the problem of forecast verification we have to define our matter of debate, that is what do we mean with the term "weather forecasts". In general we can define a forecast as *an affirmation voted to describe the future state of a system*[1]. This definition is extremely general and, for our purposes, needs some more specifications. In particular we define as *objective* those forecasts that are obtained trough objective techniques, this kind of forecasts are independent from the person or computer that are issuing them. On the contrary we will define as *subjective* those forecasts that are dependent from a subjective contribution, typically the judgment of a person. Both subjective and objective forecasts can be *quantitative*, if they can be expressed through a number/category or *qualitative* if they are expressed through an interpretable expression. Every weather forecaster in duty prepares and issues all of these different weather forecasts, even if sometimes (or very often?) unconsciously. This distinction is not purely academical, in fact different weather forecasts types have to be verified in different ways. In particular qualitative weather forecasts (e.g. "tomorrow comfortable temperatures") can be verified only in a qualitative way, as an example trough a telephonic poll, and not quantitatively, that is trough the techniques described in the following chapters. Then, when deciding to perform the weather forecast verification, the first activity that we have to carry out is to classify the forecast we are planning to examine. After this classification we can decide what kind of approach to adopt.

2 A general frame for the verification of weather forecasts

> "*If we do not pay sufficient attention to the problem of physical reality, we can easily get trapped in paradoxes*"
> (D. Ruelle - Chance and Chaos)

[1] In his book, Kendall and Stuart (1963) suggested to distinguish between "forecasts" and "predictions", being the latter based essentially on deterministic considerations while the formers are much more related on subjective and interpretable considerations. This distinction is not presently in use even if, in general, we use the expression "unpredictable" if referring to a system with too complicated or unknown laws and not the expression "unforecastable". In other words adopting the distinction suggested by Kendall, we could be legitimated to issue forecasts even for unpredictable systems.

A general frame for weather forecasts verification already exists and it is the result of the efforts of several scientists, among others it is necessary to remember the contribution of Alan Murphy, see in particular Murphy and Winkler (1987) and Murphy (1991) who collected all the previous works into a clear and well defined analytical scheme. According to the works of Murphy, there are three levels according which it is possible to verify forecasts. The first level is that of *consistency*, that is it faces the question "how well this forecast is matching what is the expected situation in the forecaster's mind?" This level of verification can be considered useless because, as written in Brooks and Doswell (1996) it is hard to know (and somehow unpleasant, we add) what exactly is in the forecasters' mind. Nevertheless it is extremely important for the forecasters try to match as much as possible the issued forecasts with their mental picture of the future weather. In fact, even unconsciously, quite often forecasters try to mould their forecasts according to the users expectations. As an example when people is "praying for rain" a forecaster, if the situation is not clear, probably will not issue a clear rain forecast, this to avoid false expectations and then complaints. This behavior, even if totally human and understandable, will bring to forecasts with a low judgment and then with a lower usefulness (in the above example the forecaster will issue 40% rain probability instead of 60%). The second level, that will occupy most of these notes, is that of *quality*, that is it faces the question "how well forecasts are matching the observations?". This level could seem the easiest and straightforward way to verify forecasts but there are several aspects that have to be fixed to obtain significant and useful results, as will be shown in the following chapters. The third and last level is that of *value*, that is it faces the question "how much money is this forecast saving (earning)?". This level is probably the most interesting for common people, but it is even the most complex to face because it depends both from judgment and quality. Moreover it depends from the characteristics of the activity for which the weather forecasts are used, in other words even if the judgment and the quality is the same, the value can be extremely different for different activities.

As we saw, quality is the fundamental level for the forecast verification and it needs the exhaustive knowledge of forecasts y_i, observations o_i and their interplay $f(y_i, o_i)$, that is the knowledge of the joint probability density function $f = f(y_i, o_i)dy_i do_i$[2]. According to Murphy and Winkler (1987) this density function can be factorized using the conditional and marginal distributions into the following two forms:

$$f(y_i, o_i) = f(y_i|o_i) \cdot f(o_i) \tag{2.1}$$

$$f(y_i, o_i) = f(o_i|y_i) \cdot f(y_i) \tag{2.2}$$

The formula (2.1) is called *likelihood-base rate factorization* because it is composed by the conditional distribution $f(y_i|o_i)$ of the y_i forecasts once the o_i observations are fixed and by the marginal distribution $f(o_i)$ of observations. The formula (2.2) is called *calibration-refinement factorization* because it is composed by the conditional distribution $f(o_i|y_i)$ of the o_i observations once the y_i forecasts are fixed and by the marginal

[2]Because weather forecasts, in general, concern discrete quantities both for predictands and predictors, the proper statistical term should be joint probability mass function.

distribution $f(y_i)$ of forecasts. The joint probability density function contains all the information and, for this reason, it fully answers to the problem of forecast verification but, unfortunately, its complexity and dimensionality makes itself almost unmanageable. For this reason it is possible to introduce several scalar values that try to reduce this complexity giving informations on the quality of the forecasts. These measurements are classifiable according to the aspect they are trying to tackle as shown in Wilks (1985).

The first class of possible scalars is called *accuracy* and it refers to the measurements that deal with the differences inside the couples measurement-forecast. As an example, the mean error $\hat{e} = \sum_i^N (o_i - y_i)$ and the squared mean error $\hat{e} = \sqrt{\sum_i^N (o_i - y_i)}$ are accuracy measurements.

The second class of measurements is that of *bias*, which deals with the differences between the mean value of observations and forecasts. This class of measurements gives us informations on how different are the forecasts from climatology. Perfect forecasts have to respect climatology, but it is fundamental to remember that unbiased forecasts could be completely wrong. Usually, when trying to forecast severe weather events, it is easy to find biased forecasts. This means that the models applied, numerical as well as conceptual, are not representing well the mechanisms we are trying to forecast.

The third class is that of *reliability*. This class gives us information concerning how well our forecasts are calibrated accordint to the observations. These measurements are obtained sorting the couples forecasts/observations $(y_i|o_i)$ according to the forecasts values and looking how well each forecast value reproduces the corresponding observation. As an example, if we want to forecast the percent probability of thunderstorms for the following day, reliability measurements are telling us if our estimation of thunderstorm probability, say 40%, corresponds exactly to the 40% observed frequency. This example will be expanded in the paragraph concerning the attributes diagram. It is evident that reliability measurements are extremely important in decision making and for all the people interested in an operational and continuous use of weather forecasts.

The fourth class is that of *resolution*, which faces the question of how well the forecasts sort observations in classes that are each other distinct. Following the example introduced above, if our forecasts of 30% and 40% thunderstorm probability for the following day both correspond to an observed frequency of 40%, then our forecasts do not resolve those two classes of observations. In other words the model adopted is not able to recognize the differences between those classes. The *resolution* measurements make use of the calibration-refinement factorization and in particular of the $f(o_i|y_i)$ conditional distribution of the joint probability density function (gives the observations probability once a forecast is fixed).

The fifth class is that of *discrimination* and it is the converse of resolution, because it faces the question of how different are the classes of forecasts sorted in function of observations. The *discrimination* measurements make use of the likelihood-base rate factorization and in particular of the $f(y_i|o_i)$ conditional distribution of the joint probability density function (gives the forecasts probability once an observation is fixed). The discrimination measurements are another way, together with those of resolution, to understand if the models used to prepare the forecasts are matching well, and how well, reality.

The sixth class is that of *sharpness* and these measurements are trying to answer to the question of how great is the variability of our forecasts, independently from the observations. This class of measurements takes care only of the marginal distribution $f(y_i)$ of forecasts and gives information on how our forecasts are spread. In general forecasts that are not spread and are almost all distributed around the climatological value have a low value of sharpness and probably are not able to reproduce the variability of the phenomena that are trying to foresee.

Facing the problem of forecast verification in the previous paragraph we introduced the joint probability density function $f = f(y_i, o_i)dy_i do_i$ of forecasts and observations. This function, whichever could be its factorization, contains the answers to all our questions, but these informations are not in a useful form because of its complexity. In other words this function is as a huge fat pig[3] when you are hungry. It can solve all your troubles but you have to know how to deal with it. The measurements, as shown on the previous sections, give us some useful, already digested, information but at the price of loosing other information. For these aspects, the distribution-oriented and the measurement-oriented approach are somehow complementary and adopting one or the other depends on our goals. In general a good compromise can be that to use a distribution oriented approach represented by several different measurements.

3 Contingency tables and related measurements

> "*We cannot assume that any condition of weather, normal or abnormal, can be ascribed to a single cause.*"
> (Sir. Napier Shaw - Manual of Meteorology Vol.I)

Contingency tables are a graphical representation of the joint probability density function in the case of discrete predictands (predictors). Being the meteorological observables usually discrete, contingency tables are extremely useful, as shown, among others, in Doswell et al. (1990) and Murphy (1996). Contingency tables are $N \times M$ tables where N is the number of discrete categories of predictors (your forecasts y_i) and M the number of discrete predictands (e.g. your observations o_i). The contingency table is composed by the relative (or percent) frequencies of all the possible couples (y_i, o_i) of forecasts and observations.

Table 1. Example of contingency table.

Forecasts/Observations	o_1	o_2	...	o_N
y_1	$f(y_1, o_1)$	$f(y_1, o_2)$	$f(y_1, ...)$	$f(y_1, o_N)$
y_2	$f(y_2, o_1)$	$f(y_2, o_2)$	$f(y_2, ...)$	$f(y_2, o_N)$
...	$f(..., o_1)$	$f(..., o_2)$	$f(..., ...)$	$f(..., o_N)$
y_N	$f(y_N, o_1)$	$f(y_N, o_2)$	$f(..., ...)$	$f(y_N, o_N)$

Contingency tables with too many categories are unmanageable, for this reason it is useful to reduce the number of categories up to two or three. As an example we can

[3]If you are vegetarian you can think to a soy bean plant

proceed in analyzing the probability thunderstorm forecasts. In this case, even if the categories can be as many as you want, it is useful to reduce them to two, that is to transform the probability forecasts in dichotomic forecasts (thunderstorm yes, thunderstorm not). The following table represents the obtained contingency table.

Table 2. Example of contingency table for dichotomous thunderstorm forecasts (yes or not).

Forecasts/Observations	T.storm observed	T.storm not observed
T.storm forecast (it should occur)	A (yes/yes)	B (yes/not)
T.storm not forecast (it should not occur)	C (not/yes)	D (not/not)

In this example, easily generalizable, perfect forecasts should have B=C=0, then all the stormy days are correctly forecast as stormy, while all the not stormy days are forecast as not stormy. Because such good situations are not common, it is necessary to retrieve from the contingency table some measurements that could help us in the verification of the quality of that forecasts. The simplest measurement introduced, which is an accuracy one (see the previous section), is the *hit rate* defined as the ratio of correct forecasts (thunderstorm forecast when it happens and thunderstorm not forecast when it does not happen) over all the forecasts issued:

$$HIT = \frac{(A+B)}{(A+B+C+D)} \tag{3.1}$$

Before to continue it is necessary to say that, just to increase confusion, *hit rate* in some texts is called *percent correct*; usually in those texts the expression *hit rate* is reserved for the *probability of detection* (see below). The hit rate spans from 0 to 1 and higher values are better than lower ones. This kind of measurement can seem straightforward but, specially in case of rare events, that is for events with a small climatic frequency, can be extremely misleading. This will be evident in the following example. Suppose to live in a place where the climatic frequency of stormy days is 0.1 (that is 10 stormy days every 100 days). In this case a forecaster, aware of climatology, that issues only 10 forecasts of stormy day but, unfortunately, in the wrong days he or she will have a quite high, even if useless, hit rate that is

$$HIT = \frac{(0+80)}{(0+10+10+80)} = 0.8$$

Better results could be obtained by a forecaster that does not take care of climatology and issues only forecasts of not stormy days. In this case these useless forecasts will obtain an higher hit rate than before, that is

$$HIT = \frac{(0+90)}{(0+0+10+90)} = 0.9$$

Trying to correct this behavior, an other measurement was introduced, that is the *threat score* (or *critical success index*) which is defined as the ratio between the number of correct forecasts of the event (e.g thunderstorm forecast and observed) and the number of thunderstorms forecast or observed

$$TS = CSI = \frac{(A)}{(A + B + C)} \tag{3.2}$$

Threat score spans from 0 to 1 and, as for the hit rate, higher values are better than lower ones. Using the threat score instead of hit rate, the performances of the forecasters in the previous examples had been both the same, that is

$$HIT = \frac{(0)}{(0 + 10 + 10)} = \frac{(0)}{(0 + 0 + 10)} = 0.0$$

which are clearly a better measure of the quality of those forecasts.

One of the troubles of this measure of quality is that it is not "equitable". The term *equitability* (or *equitable*) usually means the property of being immune from the contribution of random correct forecasts. Plain speaking this means that even random forecasts, usually, will have a TS different from zero, then they will show some forecasting skill. For this reason it could be useful to eliminate in the TS this contribution, highlighting only the "true" forecasts skill. This can be done eliminating from the number of correct forecasts (A) the number of correct forecasts I'll have only thanks to the chance. This number, using the above contingency table, is given by the formula

$$A_{chance} = (A + B) \cdot \frac{(A + C)}{(A + B + C + D)}$$

then the *equitable threat score* (ETS) will become

$$ETS = \frac{(A - A_{chance})}{((A - A_{chance}) + B + C)} \tag{3.3}$$

The ETS so defined is, by construction, equitable but unfortunately depends from the correct rejections (D in the contingency table) by way of the total number or events N.

Another measure of accuracy is the *probability of detection*, defined as the ratio between the number of correct forecasts of the event (e.g. thunderstorm forecast and observed) and the number of times the event is observed, that is

$$POD = \frac{(A)}{(A + C)} \tag{3.4}$$

As already mentioned, in some texts the *probability of detection* above described is called *hit rate*. In those texts the index here called *hit rate* is instead called *percent correct*. Pay attention! This measurement spans from 0 (no detection at all) to 1 (perfect detection) it is useful because it gives information on our ability to recognize the occurrence of the event but it can even be misleading if used alone, in fact if a forecaster, on the same climatology of the previous examples, issues alway a forecast of stormy day, he or she will have a perfect detection but a low threat score, that is

$$POD = \frac{(10)}{(10+0)} = 1$$

$$TS = \frac{(10)}{(10+90+0)} = 0.1$$

In this one can guess that the forecasters have to make exercises on recognize when thunderstorms will not take place.

Another measurement that can be useful in dealing with the ability of detection is the *false alarm rate*, that is the ratio between the number of times the event is forecast and did not take place and the number of all the times the event is forecast, that is

$$FAR = \frac{(B)}{(A+B)} \tag{3.5}$$

The *false alarm rate* spans from 0 (no false alarms at all) to 1 (every forecasts is a false alarm) and in general it is a good habit to use it every time we use the probability of detection. In the example above, the false alarm rate had been

$$FAR = \frac{(90)}{(10+90)} = 0.9$$

In some texts the FAR is called *false alarm ratio* and the expression *false alarm rate* is reserved for the quantity $(B)/(B+D)$.

The last measure that we can retrieve from the contingency table is the *bias*, defined as the ratio between the number of times an event is forecast and the number of times an event occurs, that is

$$BIAS = \frac{(A+B)}{(A+C)} \tag{3.6}$$

The bias spans from 0 (events never forecast) to infinity and it is useful because it tells us if our models are well calibrated on the climatology of the events. If we forecast more events (say stormy days) than observed, our forecasts will have a bias higher than 1 (we over-forecast), if we forecast an event less frequently than its climatology our forecasts will have a bias lower than 1 (we under-forecast). Obviously, if we are well calibrated on the climatology our bias will be exactly 1, that is then the desired value. Of course, perfect forecasts will have a $BIAS = 1$, but forecasts with a $BIAS = 1$ should not necessarily be perfect and, in some circumstances, could be even acceptable a bias higher or lower than the perfect one. In the climatic forecasts of stormy day discussed above, the bias had been

$$BIAS = \frac{(0+10)}{(0+10)} = 1$$

while in the "always yes" stormy day forecasts discussed above the bias had been

$$BIAS = \frac{(10+90)}{(10+0)} = 10$$

In general, with simple algebra, it is possible to explain all the above mentioned scores as function of other scores. Apart for a mere algebraic exercise, this aspect can be useful to interpret the trends in the forecasts performances. As an example we can write the relation that connects BIAS, POD and FAR. With simple algebra we obtain

$$BIAS = \frac{POD}{FAR} \cdot \frac{1}{(\frac{1}{FAR} - 1)} = \frac{POD}{(1 - FAR)}$$

then, applying the simple differentiation rules, we have

$$d(BIAS) = \left(\frac{POD}{(1 - FAR)}\right) \cdot \frac{d(POD)}{POD} + \left(\frac{POD}{(1 - FAR)}\right) \cdot \left(\frac{FAR}{1 - far}\right) \cdot \frac{d(FAR)}{FAR}$$

then, using again the above expression for the BIAS, we have

$$\frac{d(BIAS)}{BIAS} = \frac{d(POD)}{POD} + \frac{FAR}{(1 - FAR)} \cdot \frac{d(FAR)}{FAR} \tag{3.7}$$

This relation defines the connection between the relative changes in POD, FAR and BIAS and we can use it to interpret what shown in the Figure 1. From 1995 to 2004 the OSMER's thunderstorm forecasts show an increase of POD together with an increase of BIAS and an almost constant FAR. The above equation tells us that the increase in BIAS can be almost entirely ascribed to an increased ability in recognizing situations prone to the thunderstorm onset.

Before to conclude it is important to point our attention toward an aspect that is not marginal even if not enough stressed in the previous paragraphs, that is the fixing of the thresholds. In other words it had been arbitrary, even quite natural, to fix at the level of 50% the threshold to consider a forecast as a "stormy day forecast". Moreover no one defined when a day is a stormy one. These aspects, that can change dramatically the results of our forecasts verification, will be faced in detail in the last chapter.

4 Skill scores

> *"Physicists use statistics as drunken people use street lamps: not for their light but just to stand up"*
> (Anonymous)

Skill scores are a measure of the relative accuracy of a set of forecasts in comparison with a reference set. Skill scores are useful because they are an adimensional number that tells us if our models or abilities (skills) are better than others. To define a skill score we need an i) accuracy measurement (see the above sections) and ii) a method to prepare the reference forecasts. Then the accuracy measurement have to be computed for our forecasts, say M its value, and for the reference forecasts, say M_{ref} its value. If M_{perf} is the measurement associated to perfect forecasts, the skill score has the form

$$Skill\ Score = SS = \frac{(M - M_{ref})}{(M_{perf} - M_{ref})} \tag{4.1}$$

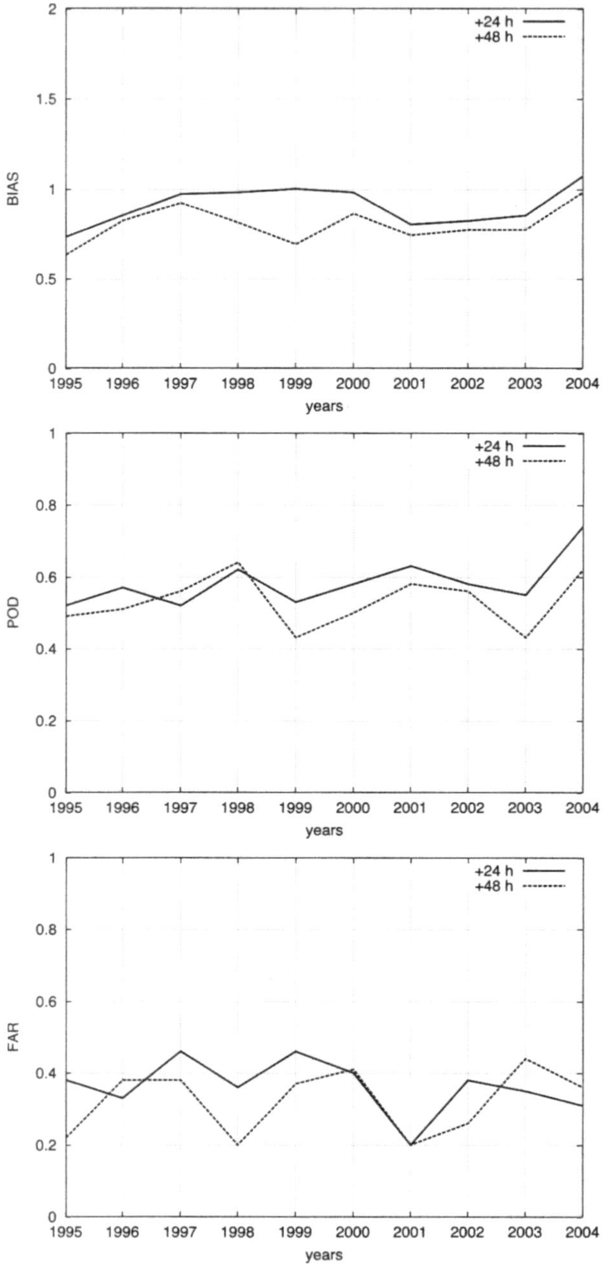

Figure 1. The time series of BIAS (upper panel), POD (central panel) and FAR (lower panel) for the dichotomous thunderstorm forecasts in the plain of Friuli Venezia Giulia issued by OSMER - ARPA.

Usually the reference forecasts adopted are random forecasts (forecasts issued by chance), climatic forecasts (forecasts issued by chance but respecting the climatology of the phenomena we are interested in) and the persistency forecasts (e.g. issue for tomorrow the observation of today). Moreover, skill scores can be used to compare new forecast methods with old ones to quantify the relative improvement.

Skill scores can have both positive and negative values, of course positive values are better than negative ones, because they are telling us that our forecasts have a better performance than the reference. A zero skill score means that our efforts are vain because the same performance could be obtained using the reference forecasts. In literature there are two kinds of skill scores that are generally used, that is the Heidke skill score (hereafter HSS) introduced by Heidke (1926) and the Kuiper skill score (hereafter KSS) introduced by Hanssen and Kuipers (1926); see for a complete reference Wilks (1985).

The Heidke skill score uses the hit rate as the accuracy measurement and random forecasts as a reference. It is important to stress that the random forecasts used for the Heidke skill score are not fully random because they have to conserve the marginal distributions of observations and forecasts of the real cases. In other words this means, for example, that if we deal with discrete forecasts (observations), the number on each class has to be maintain the same but issued randomly. In the case of dichotomous forecasts (e.g. thunderstorm yes or not) the Heidke skill score assumes a simple form that can easily be determined using the contingency table shown in the previous section and the factorization of the joint probability density function. In this case the hit rate of our forecasts is by definition $M = (A + D)$, while the hit rate of the reference forecasts is the product of the marginal distributions of forecasts and observations, that is $M_{ref} = f(f_{yes})f(o_{yes}) + f(f_{not})f(o_{not}) = (A + B)(A + C) + (C + D)(B + D)$ and the hit rate for perfect forecasts is $M_{perf} = 1$. Merging all these measurements we obtain

$$HSS = \frac{(A + D) - [(A + B)(A + C) + (C + C)(B + D)]}{1 - [(A + B)(A + C) + (C + D)(B + D)]}$$

that with simple algebra becomes

$$HSS = \frac{2 \cdot (AD - BC)}{(A + C)(C + D) + (A + B)(B + D)} \tag{4.2}$$

The Kuiper skill score, on the contrary, is obtained using as a reference forecast, random forecasts that are constrained to be unbiased, then the climatology of the events is preserved. In this case the marginal distribution of observations and forecasts have to be the same. In the case of dichotomic events, the Kuiper skill score can be easily determined using the contingency table, then we have $M = (A + D)$, the hit rate of the reference forecast is the product of the marginal distributions of forecasts and observations, that is $M_{ref} = f(o_{yes})f(o_{yes}) + f(o_{not})f(o_{not}) = (A+C)(A+C)+(B+D)(B+D)$ while the hit rate for perfect forecasts is $M_{perf} = 1$. Merging all these measurements we obtain

$$KSS = \frac{(A + D) - [(A + C)(A + C) + (B + D)(B + D)]}{1 - [(A + C)(A + C) + (B + D)(B + D)]}$$

that with some algebra becomes

$$KSS = \frac{(AD - BC)}{(A + C) + (B + D)} \tag{4.3}$$

5 Attribute Diagrams

"Torture your data long enough and they will confess to anything" (Anonymous)

The attribute diagrams can be considered a graphical representation of the decomposition of the *Brier Score*. The *Brier Score* (BS) is defined by the following formula

$$\frac{1}{N} \sum_{i}^{N} (ff_i - of_i)^2 \tag{5.1}$$

whereff_i is the forecast frequency of the event i and $of_i = 1$ if the event i is observed and $of_i = 0$ if it is not observed. In other words if a forecaster issues a 60% probability of thunderstorm occurrence for tomorrow on a specific area and thunderstorms will not occur, the contribution of tomorrow to his/her *Brier Score* will be $ff_i = 0.6$ and $of_i = 0$. By this definition it is clear that $BS = 0$ for perfect forecasts, then the higher is the BS the worst are the forecasts. The upper boundary of BS is $BS = 1$ and it can be statistically interpreted as the mean squared error of the N pairs of forecasts and observations. The *Brier Score* can be used to compute skill scores of the form

$$BSS = \frac{(BS - BS_{ref})}{(BS_{perf} - BS_{ref})} = \frac{(BS - BS_{ref})}{(0 - BS_{ref})} = 1 - \frac{BS}{BS_{ref}} \tag{5.2}$$

but it is much more useful if used in its decomposed form and in the graphical representation of this decomposition, that is in the *attribute diagram*.

Murphy (1963) showed that using the calibration refinement of the joint proability density function $f(y_i, o_i) = f(o_i|y_i)f(y_i)$, that is using the distribution of frequencies o_i of event occurrence for a given y_i forecast value, with some algebraic manipulation it is possible to write the Brier score in the form

$$BS = \frac{1}{N} \sum_{i}^{K} n_i (y_i - \hat{o}_i)^2 - \frac{1}{N} \sum_{i}^{K} n_i (\hat{o}_i - \hat{o})^2 + \hat{o}(1 - \hat{o}) \tag{5.3}$$

where $N = \sum_i^K n_i$ is the total number of event/forecast pairs written as the sum of the K sub-samples i build according to the K different possible forecast values y_i; $p(y_i) = \frac{n_i}{N}$ is the relative frequency of the forecast value y_i, $\hat{o}_i = p(o_i, y_i) = \frac{1}{n_i} \sum_m^{n_i} o_m$ is the ith value of the conditional distribution of observations, that is the relative frequency of event observations when the ith forecast value is issued; $\hat{o} = \frac{1}{N} \sum_m^N o_m$ is the relative value of the event occurrence. All the three terms on the right of the above equation have a "statistical" interpretation: the first was called by Murphy (1963) *reliability*, the second *resolution*, the third *uncertainty*. The *reliability* is a measure of how much

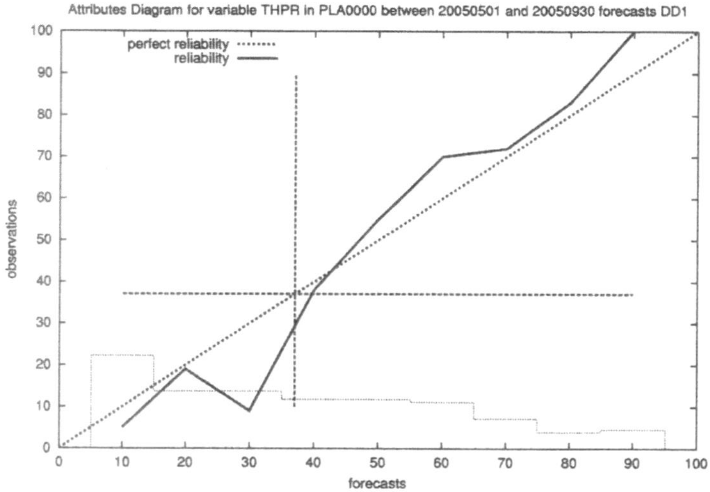

Figure 2. Example of attribute diagram obtained for the probabilistic thunderstorm forecasts in the 2005 warm season on the Friuli Venezia Giulia plain. The dotted diagonal represents the perfect calibration, while the vertical and horizontal dashed lines represent the observed thunderstorm frequency which is related to uncertainty. The histograms on the bottom of the figure represent the percent frequency of each forecast category (nine possible values, from 10% to 90% with a 10% step).

the value of the ith forecast y_i is in disagreement with the relative frequency of event observations o_i. *Reliability* is, by definition, always positive, then it has a negative effect on the BS, in fact the higher is the *reliability*, the higher (worst) is the BS [4] The second term of the BS decomposition is called by Murphy (1963) *resolution* and it is a measure of how different are the relative frequencies $\hat{o}_i$ of event observations corresponding to the ith forecast value in comparison with the climatic frequency of the event occurrence $\hat{o}$. The higher is the *resolution* the higher are the differences between the relative frequencies of event occurrence clustered according to the K different forecast values, in other words if a forecaster has a high resolution he/she is enough skillful to recognize the difference between different meteorological situations. *Resolution* is, by definition, always positive and, thanks to the "minus" sign on the formula (5.3), it has a positive effect on the BS. Both reliability and resolution are function of the forecasts (and then of the forecasters' behavior), *reliability* explicitly, *resolution* implicitly, then forecasters can at least try

[4]Because the higher is the *reliability* component of the BS the worst is the agreement between the forecast value and the corresponding frequency of event observations, it should be more appropriated to call it *unreliability*.

to do something to reduce *reliability* and increase *resolution*. The third and last term of the BS decomposition is called by Murphy *uncertainty*. It is a function only of the climatic frequency of event occurrence $\hat{o}$. Uncertainty is, by definition, positive defined, then it has always a negative effect on the BS. The highest, that is worst, value of uncertainty corresponds to the climatic frequency $\hat{o} = 0.5$. This fact can be naively understood thinking that if an event has a climatic frequency $\hat{o} = 0.5$, forecasters will not have neither the small advantage of issuing a climatic forecast (e.g. say "yes" if the event usually occurs and "not" if it usually does not occur) when they do not know what to do. The above mentioned aspects can be easily represented in a diagram called "Attribute Diagram" introduced by Hsu and Murphy (1986) for probabilistic forecasts. This Attribute Diagram shows on the abscissa the classes of probability forecasts y_i, while in ordinates it shows the value of the frequency of event occurrence corresponding to each probability forecasts y_i. If forecasts are well calibrated, then the frequency of event observations corresponding to each probability forecasts y_i should stay almost on the bisecting line of the Attribute Diagram. The Attribute Diagram reports even the frequency threshold represented by the climatic event frequency. If the points on the Attribute Diagram corresponding to the event frequencies of each y_i probability forecast are nearer to the $\hat{o}$ threshold, then the corresponding $i-th$ forecast value gives a positive contribution to the BS. An example of Attribute Diagram is shown in Figure 2 where the perforances in the probabilistic thunderstorm forecasts on the Friuli Venezia Giulia plain are shown. From Figure 2 it is clear that the overall probabilistic thunderstorm forecasts are well calibrated apart from the 30% category which is underestimated (only 10% of the days with a 30% thunderstorm probability forecast hosted a thunderstorm).

The interesting aspect of the *Attribute Diagram* is that it can be easily extended from probabilistic forecasts to physical forecasts (e.g. amount of rain, temperature, etc.) with the same useful interpretation of the reliability, resolution and uncertainty and with the only difference represented by the fact that you are dealing with dimensional values. $\hat{o}_i$ in the latter case becomes the average value of observations associated to the $i - th$ forecast, while $\hat{o}$ is the average value of the observations. If the predictand is categorical or discrete (e.g. rain amount is usually divided in classes) the extension is immediate, if this is not the case (e.g. minimum or maximum temperatures) then the predictand should be divided in classes. Even in the latter case, in general, there are no problems at all because usually forecasts are not continue but discrete (e.g. maximum temperature forecasts are expressed in terms of tens of ^oF or in a range of a few ^oC). An example of Attribute Diagram obtained for categorical forecasts is shown in Figure 3 where the daily rain amount forecasts in the 2005 warm season on the Friuli Venezia Giulia plain are represented. From this figure it is clear that the category 5-10 mm/day is underestimate (when 5-10 mm/day of rain are forecast usually you can expect 10-30 mm/day) while the category 30-60 mm/day is overestimated (when 30-60 mm/day of rain are forecast usually you can expect 10-30 mm/day).

In both the standard and extended version of the Attribute Diagram it is worth and useful accompany the diagram itself with an ancillary representation of the number (or frequency) of each class of forecasts. These extra information can be extremely useful for the interpretation of the results. In particular it is possible to interpret the fluctuations of the $\hat{o}_i$ values around the best calibration line (e.g. the bisecting line of the Attribute

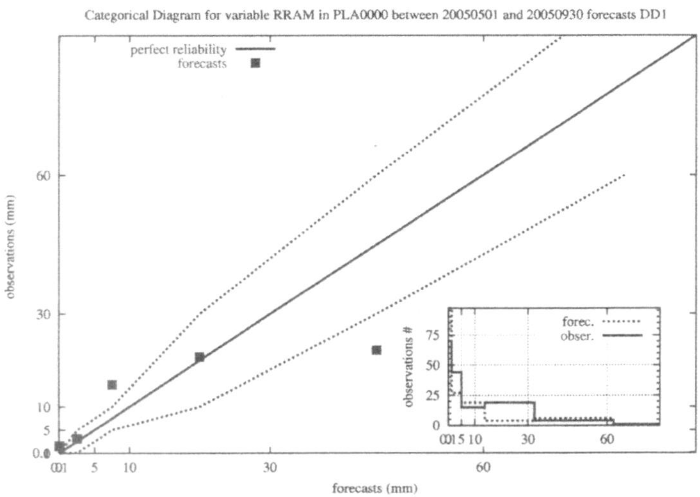

Figure 3. Example of attribute diagram obtained for the daily rain amount forecasts in the 2005 warm season on the Friuli Venezia Giulia plain. Solid diagonal represents the perfect calibration while the dotted lines represent the range of each category of rain amount forecasts. Filled squares represent the observed average values associated to each rain amount category.

Diagram). If the number of observations in a class of forecasts is low (say of the order of 10 or lower) then fluctuations can be ascribed to mere chance, but if the number of observations is higher than the fluctuation, it can be considered as a lack in the ability of distinguishing between contiguous forecast values. This can be simply described with this simple example. If we are dealing with the probability of thunderstorm occurrence in the next day and we are "well calibrated forecasters" then the 40% value of the forecast y_{40} will correspond to an event observation frequency of the order $o(y_{40}) \sim N_{40}\frac{\hat{o}_{40}}{N_{40}} = \hat{o}_{40}$, where N_40 is the number of times the 40% forecast is issued and $\hat{o}_{40}$ is the empirical event observation frequency for the y_{40} class of forecasts. If we are "not well calibrated forecasters" and, as an example, we are not able to distinguish between situations that should be classified as 40%, 50% and 60%, then the event observation frequency for the class of forecasts y_{40} will be of the order of

$$o(y_{40}) \sim \left(\frac{N_{40}}{3}\frac{\hat{o}_{40}}{N_{40}}\right) + \left(\frac{N_{40}}{3}\frac{\hat{o}_{50}}{N_{40}}\right) + \left(\frac{N_{40}}{3}\frac{\hat{o}_{60}}{N_{40}}\right) > \hat{o}_{40}$$

with similar considerations it is easy to see that under the above mentioned assumptions you will obtain $o(y_{50}) \sim \hat{o}_{50}$ and $o(y_{60}) < \hat{o}_{60}$.

6 ROC curves

"*A humpbacked is happy when sees a bigger hump*"
(Yiddish proverb.)

Sometimes, specially dealing with objective weather forecasts (typically indexes of something), see among others Manzato (2003) of dichotomous events (yes/not), we have to keep into account a continuous predictor. Different values of the predictor will have different scores (different HIT, POD, FAR, etc.) then it could be difficult to quantify the differences between various types of predictors to chose the best one. To face this problem it has been developed a graphical tool called Relative (or Receiver) Operating Characteristic curve (or simply ROC curve). This curve reports, for several values of a chosen kind of predictor, the probability of detection $POD = \frac{A}{A+C}$ as a function of the frequency of non occurred events forecast as occurring, quantity that sometimes is called Probability of False Detections $POFD = \frac{B}{B+D}$ [5]. For a monotonically growing predictor (e.g. the higher the index, the higher the probability or intensity of the event) what is expected to happen is that, changing the threshold to consider "yes" the forecast, low values of the threshold should produce high values of POD and $POFD$, while high values of the threshold should produce low values of both POD and $POFD$. In fact, adopting a low value of the threshold, the predictor will issue a small number of "yes" forecasts, then the predictor will "catch" a small number of occurring events (small POD - a not good characteristic for a predictor-) but even make a small number of mistakes forecasting events that do not occur in reality (small $POFD$ - a good characteristic of the predictor-). What is expected for a good predictor is that, increasing the threshold, POF will grow faster than $POFD$, then the ROC curve should have a negative second derivative.

With the computing of the empirical ROC curves for two or more different predictors it is then possible to compare their performances in forecasting the same class of events. In general a predictor whose ROC curve has a bigger hump, then it is staying always above the ROC curve of another predictor, it can be considered intrinsically better than that, no matter the threshold adopted. This comparison can be made graphically and it is generally very useful. As an example in Figure 4 are shown the ROC curves obtained for two different thunderstorm occurrence indexes, the EV and COIN indexes Giaiotti and Stel (2001). In that figure it is clear that the overall performance of the COIN index, indipendently from the threshold adopted, is better than that of EV index, in fact the probability of false detection of COIN index grows slowly than its probability of detection.

An attempt to make it automatic was made computing the integral of the ROC curve, that is its area. In this way it is possible to obtain a scalar measurement for each type of predictors to make the comparison and then quantify the differences. This scalar value has some evident advantages (among others it is not subjective as the graphical comparison) but as the graphical comparison does not solve some tricky cases where

[5] In some literature the Probability of False Detection is called False Alarm Rate. In those texts, just to augment confusion, what here is called the False Alarm Rate is instead called False Alarm Ratio.

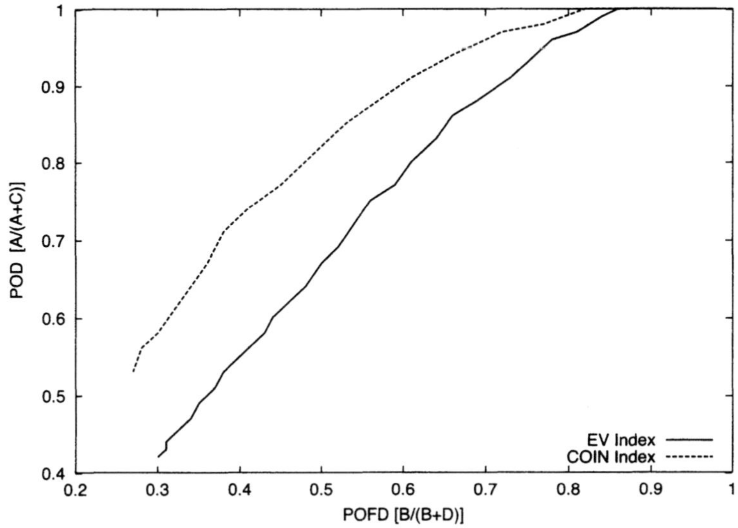

Figure 4. ROC curves computed for two different predictors, EV and COIN indexes, of thunderstorm occurrence on the plain of Friuli Venezia Giulia. The two predictors have values that differ both in range and in amplitude, nevertheless it is clear that COIN index is characterized by a ROC curve that is better shaped than that of EV index.

ROC curves of different predictors have somewhere one or more overlapping points.

Recently, making some assumptions, the use of the ROC curve was extended to subjective tornado forecasts by Brooks (2004) and Glahn (2005), demonstrating the power of this simple method in making comparisons between different kind of predictors and forecasts.

7 Verification of numerical models

"Consider anything, only don't cry."
(L. Carroll - Through the looking-glass.)

Numerical models are essential instruments for the modern weather forecasts, specially in dealing with long term weather forecasts. For this reason it is important try to verify their performances. The amount of information forecast by numerical models, that is their dimensionality, is generally higher than that of human forecasters, for this reason the verification of numerical models is, if possible, even a more complicated and tricky field, as shown among others in De Elia and Laprise (2003). The general frame of numerical models forecast verification can be divided in two main branches, according to the way in which we are trying to reduce the dimensionality. The first possibility is that to verify numerical models punctually, that is comparing the point observations with the numerical model results (point verification). The second possibility is that of comparing the observed fields and their properties with the output of the numerical models (field verification).

The point verification can be faced with the same tools developed for the subjective weather forecasts, that is using contingency tables, skill scores, etc. The only complication comes from the definition of observables (see the section: The definition of reality.). The field verification of numerical models, on the contrary, is a difficult task and, so far, it still is a source of unresolved or unsatisfactorily resolved problems, specially when dealing with the deep moist convection aspects. A way classically proposed for the verification of fields is the root mean squared error (hereafter RMSE) which is defined by the simple equation

$$RMSE = \sqrt{\frac{1}{N}\sum_{p=1}^{N}\left(f_p - o_p\right)^2} \tag{7.1}$$

where the subscripts p spans on all the grid points which define the field. RMSE gives an idea of how much the forecast field averagely differs from the observed one. In other words the RMSE can be considered as the typical error of the numerical model forecasts. Other indexes were proposed to give an estimate of this aspect, among others the mean squared error $MSE = \frac{1}{N}\sum_{p=1}^{N}\left(f_p - o_p\right)^2$ and the mean absolute error $MAE = \sum_{p=1}^{N}\left|f_p - o_p\right|$. All these estimates can be used to compute the model skill score, as an example using as reference forecasts the climatology or the persistence, but they are unsatisfactory for several reasons, specially if dealing with scattered fields like the rain amount. In this case, in fact, the field usually has several zero values and only a few non zero values. For this reason it is possible to obtain low (e.g. good) RMSE (or MAE, etc.) even if the numerical model has a terrible skill in rain forecast or does not forecast rain at all. It is possible to find pathological situations in which a numerical model that does not forecast rain at all has lower RMSE that a numerical model that tries to forecast it. Another attempt, that historically was thought earlier than the RMSE and his brothers, is the S1 score. This score was developed to try to evaluate the ability of numerical models in reproducing gradients. The S1 score, see Figure 5, is defined using this formula

$$S1 = \frac{\sum_{i\in adiacent\ pairs}\left|(f_i - f_{i+1}) - (o_i - o_{i+1})\right|}{\sum_{i\in adiacent\ pairs} Max\left(\left|(f_i - f_{i+1})\right|; \left|(o_i - o_{i+1})\right|\right)} \tag{7.2}$$

Even the S1 score is unsatisfactory for several reasons, among these is the fact that differently from the RMSE it does not care about absolute magnitude of fields, then it does not care of any systematic bias. Another unappealing aspect of S1 is that it is prone to seasonality. In fact, in Spring and Autumn usually fields naturally show wider gradients, then in these two seasons the denominator of the S1 definition should be generally greater than in the two remaining seasons. For this reason Spring and Autumn S1 values are naturally lower than in the remaining two seasons. Another characteristic of S1 (and almost all of the others similar indexes) is that its amplitude is domain dependent, then the same numerical model verified using the same observed field could obtain different scores only widening or shrinking its domain or changing the characteristics of its grid. This is somehow obvious, even under the conceptual point of

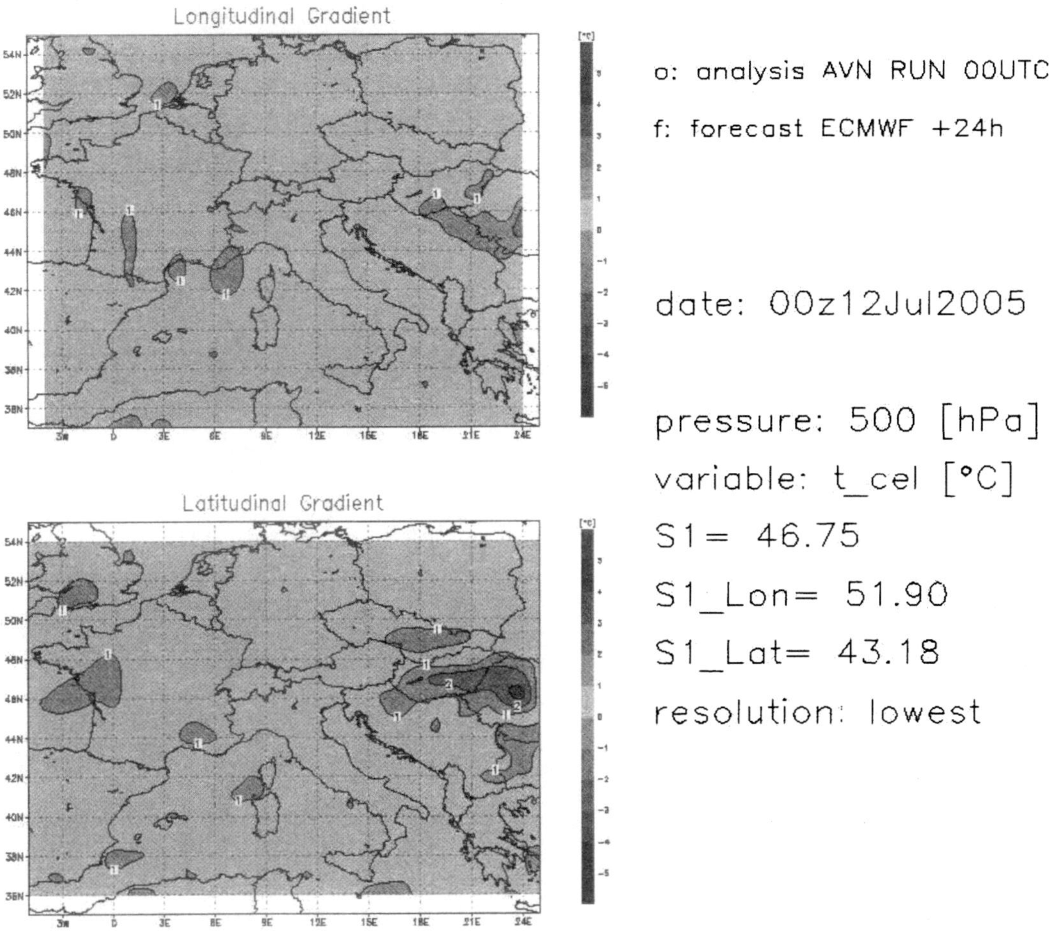

Figure 5. Example of S1 score.

view, because what we can do is to verify the quality of the fields and not the quality of the numerical model in itself. Currently, to the authors knowledge, the S1 score is almost unused.

Another commonly used score used for the verification of fields is the Anomaly Correlation (hereafter AC). AC is an adimensional estimate of the average difference between the forecast and observed values on all the field grid points once they had been normalized with the climatic value of the field in each specific grid point. In formulae the anomaly correlation is defined as

$$AC = \frac{\sum_i^N (f_i - C_i) \cdot (o_i - C_i)}{\sqrt{\sum_i^N (f_i - C_i)^2 \cdot (o_i - C_i)^2}} \tag{7.3}$$

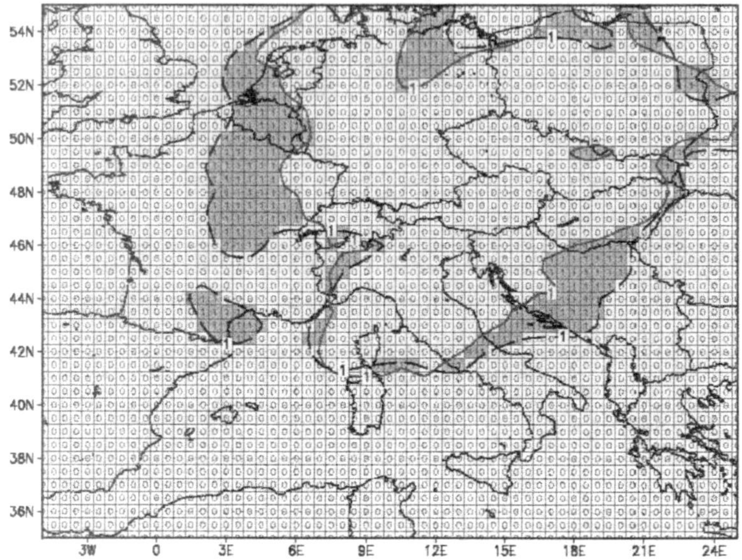

Figure 6. Example of metric adopted to measure the distance between the t=1 ^oC isopleths at the 700 hPa isobaric level. The forecast field is that obtained by the ECMWF numerical model, while the analysis used as "reality" is that of NCEP. The shaded area represents geometrically the measure of the distance.

where N is the number of field grid points and $C_i = \sum_{k=1}^{M}(o_k(i))$ is the average value of the field in the ith grid point computed using M previous fields. As above, y_i and o_i are respectively the forecast and observed value in the ith grid point. The anomaly correlation has some analogies with the Pearson's correlation coefficient but even some important differences. The analogy is that both Pearson's correlation and anomaly correlation are bounded between -1 and $+1$. The former value is reached when forecasts and observations are always (e.g. for all the grid point) in opposition of phase according to the climatology, namely when forecasts are above the mean value observations are below and vice versa. The latter value is obtained when forecasts and observations are always in phase. Unfortunately it is not possible to consider the AC a sort of correlation between forecast and observed anomalies apart for very specific cases. In fact one important difference between AC and the Pearson's correlation coefficient is that the climatic value C_i changes from point to point and it is not constant as required by the Pearson's correlation coefficient. If, as an example, C_i is constantly equal to zero, as can happen if the fields verified are quite wide both in time and space, then the similarity between Pearson's coefficient and AC can be applied. By construction AC is weakly sensitive to the forecasts bias (likely the S1 but differently from the RMSE) but it is sensitive to systematic positive or negative differences between forecasts and observations. Several studies showed that even if RMSE (or one of its brothers) and AC focus their attention on different aspects of the comparison between fields forecast and observed, then the amount of information they give is somehow complemental. Both empiric and theoretical works showed that values of AC of the order of 0.5 or below mean that the forecasts are essentially useless.

Another attempt, recently developed by the authors for the verification of fields which seems quite promising is that of focusing the attention on the most significant isopleths of the field that has to be verified, introducing a metric to measure the difference between the specified isopleths (see Figure 6). This approach bases its strength on the fact that the dimensionality of the field is reduced taking into account the isopleths and that it is potentially near both to the work of forecasters and of the numerical models developers. Moreover, with an appropriate choice of the metric used to compute the distance between isopleths, this method can deal even with scattered and noisy fields, like that of precipitation. The metrics so far adopted and actually under test are: i) the complementary of the intersection of the areas encompassed by forecast and observed isopleths; ii) the maximum of the distances between all the points forming the isopleth observed and forecast [6]. The advantage of both the above defined metrics is that they have physical dimension, moreover they can be used in a skill score using, as an example, the persistency as a reference forecast. In particular the metric i) is easily obtainable using some widespread softwares for the display of gridded data (GrADS among others) while the metric ii), apart from the measure itself, can give even the point where the maximum distance is obtained, then can give useful hints both to forecasters and model developers on some systematic effects, as an example, due to geography or orography.

8 The economical value of weather forecasts

"*How do you expect to get your money,*
if you know nothing of Meteorology"
(Aristophanes - The Clouds)

Weather forecasts are useful and can be used in several fields and situations. It is possible to quantify this usefulness in terms of costs and losses related to the decision of carrying out or not some actions. In this way it is even possible to define the amount of money saved (or not lost) following those weather forecasts. That amount of money can then be considered as the maximum value of those weather forecasts. In fact no one would like to spend more money than he or she save with those information, apart from the marginal value that will be quickly described at the end of the section. Another aspect that, hopefully, will become evident after this section is that the economical value can change in a dramatic way accordingly to the context in which it is applied, as shown by Roebber and Bosart (1996) among others, and according to the costs and losses related to the specific activity that make use of the weather forecasts.

The general frame to calculate the economical value of weather forecasts, developed among others by Katz and Murphy (1997) supposes that the decision of carrying out an activity as a consequence of a weather forecast entail a Cost and that the decision of not carrying out any action entails a Loss if the event object of the forecast occurs.

[6] By definition the distance between a point and a curve is the minimum geometric distance between that point and the curve.

An example can be that of organizing a show indoor (Cost) because of a thunderstorm forecast. If the thunderstorm does not occur, then the organizers will have a Cost but if the thunderstorm will occur and the organizers did not take it into account, the result is a Loss. For these dichotomous forecasts it is possible to define a matrix, called the cost-loss matrix similar to the contingency table above defined, that is

Table 3. Contingency table for dichotomous events (yes or not) and related cost and loss.

Forecasts/Observations	Event observed (yes)	Event not observed (not)
Event forecast (it should occur - yes)	Cost A (yes/yes)	Cost B (yes/not)
Event not forecast (it should not occur - not)	Loss A (yes/yes)	Nor cost neither loss (0) D (not/not)

A person or company that uses weather forecasts to take decisions will experience an outlay or expenditure E function of the above forecasts' contingency table, that is

$$E_{forecaster} = (No.\ of\ f.cast\ occurences) \cdot Cost + (No.\ of\ not\ f.cast\ occurrences) \cdot Loss$$

then

$$E_{forecaster} = A \cdot Cost + B \cdot Cost + C \cdot Loss \tag{8.1}$$

A person or company that respectively: i) does not use any kind of action (braveheart attitude); ii) always carry out the action (coward attitude); iii) use climatic forecasts; will have, in the same situation of the above contingency table, the following expenditures

$$E_{braveheart} = (No.\ of\ occurrences) \cdot Loss = A \cdot Loss + C \cdot Loss \tag{8.2}$$

$$E_{coward} = (No.\ of\ forecasts) \cdot Cost = A \cdot Cost + B \cdot Cost + C \cdot Cost + D \cdot Cost \tag{8.3}$$

$$E_{climate} = A \cdot Cost + C \cdot Cost + \left((A+C) - \frac{(A+C)}{N} \cdot (A+C)\right) \cdot Loss \tag{8.4}$$

The difference between each pair of the above mentioned four economical expenditures represents the maximum relative value of each pair of attitudes. As an example the economical value of the "braveheart attitude" against the "coward attitude" is given by

$$E_{braveheart} - E_{coward} = \left(A \cdot Loss + C \cdot Loss\right) - \left(A \cdot Cost + B \cdot Cost + C \cdot Cost + D \cdot Cost\right) \tag{8.5}$$

where M is the marginal utility of being "bravehart" in spite of being "coward". Marginal utility is an important and powerful concept developed by social sciences that makes possible the interpretation of several strange even if common human behaviors and actions. The "braveheart vs coward" example can be useful in explain this. In fact sometimes, even if it is economically worth to be a braveheart (small Loss or high Cost), people chose to be a coward just not to live with the anxiety of decide what and when to do. This example fits quite well for the weather forecasts, when people have to justify something or took some decisions or sometimes just put the blame on someone else.

As mentioned before the economical value of a weather forecast is not possible to define without a comparison with other weather forecasts (in this case even the attitude "no forecasts at all" is considered as a kind of forecast). Moreover, even if we consider two different types of forecasts, the economical value depends from the absolute value of Cost and Loss. This means that the same weather forecast can be economically convenient for a particular activity but not convenient for another one. Nevertheless in a set of specific cases it is possible to develop a general treatment of the problem of the definition of a set of prototypical economical value of weather forecasts. This generalization, in particular, it is possible when i) we have dichotomous events (yes/not) with a climatic probability of occurrence $P_{climate}$; ii) we are dealing with weather forecasts with a probability of correct detection F_{yy} (probability of event forecast and observed) and of false detection F_{yn} (probability of event forecast but not observed) respectively higher and lower than $P_{climate}$; ii) for all the different weather forecasts sets, the respective F_{yy} and F_{yn} must be always each other higher and lower ("onion shaped" set of weather forecasts). This can be easily represented with the following equation

$$0 < F_{yn}^{a} \le F_{yn}^{b} \le P_{climate} \le F_{yy}^{b} \le F_{yy}^{a} < 1 \tag{8.6}$$

where the two superscripts a and b are referring to two different set of weather forecasts. If the three hypothesis are respected, then it is possible to define a quantity called *quality* which has the following form

$$q = \frac{F_{yy} - P_{climate}}{1 - P_{climate}} \tag{8.7}$$

which has the characteristic to be non-decreasingly linked to the economical value of the weather forecasts.

9 The definition of reality

"What is truth?"
(Pontius Pilate - Gospel of Luke)

Every forecast, to be considered scientific, must specify clearly what will be the state of the variable that we want to forecast. This is possible only if the variable which is the "observable" of the physical reality we are going to forecast is defined in an exhaustive and mutually exclusive way. In simple words this will be possible only if, whatever the physical state of the atmosphere, we will have one (exhaustivity) and only one (mutual exclusivity) possible value of the variable. The choice of the most appropriate variable (observable) is, sometimes, a tricky task for two main reasons: i) we are issuing weather forecasts for people, then we need to choose a variable that could have a meaning for people; ii) the reality we have to deal with is complex and it is hard to shrink all the aspects in a simple number. These two reasons could emerge clearly using two examples.

Thunderstorm are defined in the Glossary of Meteorology edited by Huschke (1959) American Meteorological Society as "... a local storm ... always accompanied by lightning and thunder, usually with strong gusts of wind, heavy rain, and sometimes with hail. ...". This definition is quite clear, even according to a physicist's standard[7] , also if there are some mentioned aspects (e.g. hail, gusty wind, etc.) that do not help in finding a variable that is exhaustive and mutually exclusive. In finding a right variable the attention should be pointed toward the "thunder" aspect, then toward the "lightning" aspect. This way is clearly a promising one, also because there are several automatic lightning detection systems that can give information concerning lightning and then on thunders. Unfortunately lightning detected by these systems are usually only those cloud-to-ground (hereafter CG), which are the minority of all flashes produced by thunderstorms. If we are interested in forecasting thundery days on a specific area, logic alone will suggest to adopt a definition of a dichotomous variable according which the day is thundery if on the area we expect at least one CG lightning. But is this the best choice? In Friuli Venezia Giulia forecasters adopted a different and a little bit more complex definition for several reasons. Here below these reasons are expressed and explained just to show that if we want to issue forecasts and then compare them with other forecasts, we need to keep well in mind what we are talking about.

Reason 1 (Homogeneity with the past). In the Friulian plain from 1974 to 1981 was active a network of volunteers distributed with a grid mesh of nearly 4 km. These volunteers collected reports concerning thunderstorms (at least one thunder heard during the day) and hail (at least an hailstone with a diameter wider than 0.5 mm).

Reason 2 (Characteristics of the modern lightning detection system available). The system actually working in Italy has a nominal precision of nearly 600 m (one standard deviation), then we consider that a CG lightning position is nominally correct with an error of nearly 2 km.

Reason 3 (People's sensitivity). Usually people in Friuli Venezia Giulia considers a day as thundery if there is at least rain associated to it. Moreover issuing weather forecasts on a wide area, forecasters were interested in relatively widespread (for the scale of the plain, e.g. 5000 km^2) phenomena.

Merging the three above reasons forecasters and forecasts verifiers decided to divide the plain and coast of Friuli Venezia Giulia according to a network with mesh size of 4 km and to define a day as thundery if there were at least three 4×4 boxes hit by a CG lightning. In this way it was possible to maintain a sort of homogeneity (at least climatic) with the past (Reason 1); to avoid the effects of instrumental errors (Reason 2); to point the forecasters attention and efforts only in events significantly widespread,

[7]Unfortunately not all the definitions generally adopted in meteorology are clear like this. An example is the case of "mesocyclone", for which are available several and in some case contradictory definitions.

e.g. wider than 1% of the whole forecasting area and at least with some rain statistically reported on the same area (Reason 3). This definition is clear, exhaustive and mutually exclusive, nevertheless it is evident that different definitions should be possible and they might have different performances.

Presently, because of the evolution in weather forecasts, since in Friuli Venezia Giulia thunderstorm forecasts are issued even for areas smaller than the plain (e.g. for the major towns) the definition adopted for these small scale forecasts is different from that above described. In particular a day is considered as thundery if there is at least a CG lightning in an area within a circle with a diameter of 20 km centered on the center of the specific area. This choice was done because it maintains homogeneity with the past (Reason 1)[8] and avoids troubles related to instrumental errors (Reason 2). Other definitions might be possible but the important aspect, probably never enough stressed, is that every forecast and forecast verification should start with a precise definition of predicand and predictor. Particular attention have to be given to the area took into account for the choice of definition.

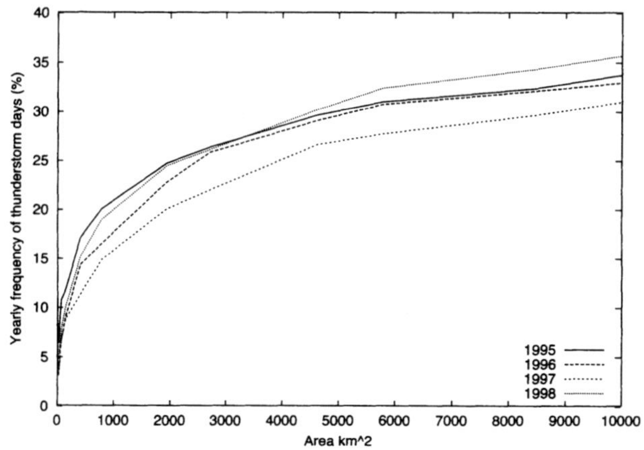

Figure 7. Yearly frequency of thunderstorm days as a function of the area taken into account in the years from 1995 to 1998. In this case a day is considere as "thunderstormy" if there is at least one cloud-to-ground lightning observed in the 24 hours on the fixed area.

In figure 7 it is shown the observed frequency of thunderstory days as a function of the area used in the defnition and in the years from 1995 to 1998. The observd frequency increases monotonically with the area itself and this fact, apart from its intrinsic interest (it can be shown that the "knee" of the curve is related to the average area covered by a thunder cell) it tells us that the quality of our thunderstorm forecasts are a function of the

[8]Remember that thunderstorms can be heard only if they are nearly within 20 km from the observer. With the above definition every person within a circle of 20 km centered on the town's center has the possibility to hear the thunder. In this way it is possible to maintain homogeneity with the past observations historically carried out in the towns and which considered a day as thundery if there was at least a thunder heared.

area fixed in our definition of thunderstorm day. For a given set of thunderstorm forecasts, just increasing the area of the "sensitive surface" for the thunderstorm definition, we can decrease our FAR and increase our POD, then increase or decrease the quality of the forcasts. This can be quite easily shown for purely random dichotomous forecasts (i.e. no correlation between forecasts and observations), in fact in that case

$$POD = P_{climate}$$

$$FAR = (1 - P_{climate})$$

while

$$TS = \frac{P_{climate}}{\left(1 - P_{climate} + \frac{P_{climate}}{P_{yes}}\right)}$$

were P_{yes} is the frequency of *yes* forecasts and $P_{climate}$ is the climatic frequency of the event [9].

The definition of reality can be a complicated task when dealing with the winds as well. In that case the variable is characterized not only by a numerical value (modulus) but even by a direction, moreover its intrinsic variability reduces the meaning of the average procedure. In the case of maximum wind speed forecasts, sometimes issued in case of severe weather events, attention can be given essentially toward the wind intensity, avoiding the troubles connected with the direction and with the variability of the wind that, in this case, can be treated as a scalar observable. For a general point of view, then even in case of weak wind fields, important for the dispersion of pollutants and/or for touristic purposes, a general solution has not been developed up to now. A possible approach is that to stratify the wind measurements available for the period object of the forecast (e.g. 24 hours, 6 hours, etc.) in octants (or quadrants) according to their direction. The most frequent octant is assumed as the characteristic wind of the period, with a modulus equal to the average wind speed for that octant. If two octants have the same frequency and are contiguous, it is assumed that the characteristic wind is that of the "quickest" octant. If the frequency and average velocity of two octants is the same or the two octants are not contiguous, then the wind is considered as *variable* in the specific period. This scheme has been developed keeping into account the fact that usually the main winds are defined according to their direction (i.e. the compass or wind rose).

Dealing with numerical models, then with fields, the situation is even more complex. In fact if we are interested in verifying the quality of a numerical model (both from the punctual or field point of view) we have to keep into account even its time and spatial resolution. Numerical models are essentially tools that solve differential equations using finite differences, then their outputs are essentially the result of a volume average. Dealing with smooth fields, like geopotential, the average process will not change dramatically things. Dealing with scattered fields (like rain) the results can change completely because low resolution numerical models will intrinsically produce smoother results than high resolution ones. For this reason comparing the output retrieved on a grid point by

[9] The condition of randomness is needed to write the contingency table using the information coming only from the event frequency and the forecast frequency without their covariance

a numerical model with a rain gage measurement on the same point is intrinsically a mistake because nature works on the basis of a continuum, while the numerical model works on the basis of a grid. In this case the solution is to use several rain gages (if possible) and compute the average on a grid which matches that of the numerical model.

Before to conclude, it is important to stress that the average procedure is not always a good choice as is in comparing numerical models with point observations. In evaluating severe weather forecasts, as for example heavy rain, when issuing a "heavy rain warning" we are pointing our attention on the extreme value. For this reason, even if we have in the same area several rain gages, the average procedure of their measurement to define the observable for "heavy rain" could bring to an underestimate of the event itself. In this case the choice of the maximum value should be, probably, a better one.

Bibliography

L. F. Bosart. Whither the Weather Analysis and Forecasting Process?. *Wea. and Forecasting*, 18:520–529, 2003.

H. E. Brooks and C. A. Doswell. A comparison of measure-oriented and distribution-oriented approaches to forecast verification. *Wea. and Forecasting*, 11:288–303, 1996.

H. E. Brooks. Tornado-Warning Performance in the Past and Future. *Bull. Amer. Meteor. Soc.*, 85:837–843, 2004.

R. De Elia and R. Laprise. Distribution-Oriented Verification of Limited-Area Model Forecasts in a Perfect-Model Framework. *Mon. Wea. Rev.*, 131:2492–2509, 2003.

C. A. Doswell III, R. Davies-Jones and D. L. Keller. Trends in the Quality of National Weather Service Forecasts. *Wea. Forecasting*, 1:42–55, 1986.

D. Giaiotti and F. Stel. A comparison between subjective and objective thunderstorm forecasts. *Atmos. Res.*, 56:111–126, 2001.

B. Glahn. Tornado-Warning Performance in the Past and Future-Another Perspective. *Bull. Amer. Meteor. Soc.*, 86:1135–1141, 2005.

A. W. Hanssen and W. J. A. Kuipers On the relationship between the frequency of rain and various meteorological parameters. *Meded. Verh.*, 81:2–15, 1965.

P. Heidke. Berechnung ds Erfolges und der Güte der Windstärkevorhersangen im Sturmwarnungdienst. *Geogr. Ann.*, 8:310–349, 1926.

W. R. Hsu and A. H. Murphy. The attributes diagram: a geometric framework for assessing the quality of probobility forecasts. *Int. J. Forecasting*, 2:285–295, 1986.

R. E. Huschke. Glossary of Meteorology. Amer. Meteor. Soc. Pub. 1989.

R. W. Katz and A. H. Murphy. Economical Value of Weather and Climate Forcasts. Cambridge Univ. Press, 1997.

M. G. Kendall and A. Stuart. *Advanced theory of statistics. v.1: Distribution theory.* 2nd ed. Griffin, 1963.

F. Le Blancq and P. Johnson. Will poor forecasts spoil your holiday?. *Weather*, 58: 203–211, 2003.

A. Manzato. A climatology of instability indices derived from Friuli Venezia Giulia soundings, using three different methods. *Atmos. Res.*, 67-68:417–454, 2003.

A. H. Murphy. A new vector partition of probability score. *J*. Appl. Meteor., 12:595–600, 1973.

A. H. Murphy. Forecast verification: Its Complexity and Dimensionality. *M*on. Wea. Rev., 119:1590–1601, 1991.

A. H. Murphy. The Finley Affair: A signal Event in the History of Forecast Verification. *W*ea. Forecasting, 11:1–20, 1996.

A. H. Murphy and T. E. Sabin. On Summary Measures of Skill in Rare Event Forecasting Based on Contingency Tables. *W*ea. Forecasting, 5:576–585, 1990.

A. H. Murphy and R. L. Winkler. A general framework for forecast verification. *M*onthly Weather Review, 115:1330–1338, 1987.

P. J. Roebber and L. F. Bosart. The Complex Relationship between Forecast Skill and Forecast Value: A Real-World Analysis. *W*ea. Forecasting, 11:544–559, 1996.

M. D. Vescio and R. L. Thompson. Subjective Tornado Probability Forecasts in Severe Weather Watches. *W*ea. Forecasting, 16:192–195, 2001.

D. S. Wilks. *S*tatistical Methods in the Atmospheric Sciences. Academic Press, 1985.

Index

-M-

-N-

-O-

-P-

-Q-

-R-

-S-

-T-

-U-

-V-

-W-

-Y-

-Z-

FSC
www.fsc.org
MIX
Papier | Fördert gute Waldnutzung
FSC® C083411

Kolibri 360 GmbH
Ferdinand-Jühlke-Straße 7
99095 Erfurt, Deutschland
produktsicherheit@kolibri360.de